Supply
Chain
Management

Supply Chain Management

SAMUEL H. HUANG

CRC Press
Taylor & Francis Group
Boca Raton   London   New York

CRC Press is an imprint of the
Taylor & Francis Group, an **informa** business

CRC Press
Taylor & Francis Group
6000 Broken Sound Parkway NW, Suite 300
Boca Raton, FL 33487-2742

© 2013 by Taylor & Francis Group, LLC
CRC Press is an imprint of Taylor & Francis Group, an Informa business

No claim to original U.S. Government works

Version Date: 20130515

International Standard Book Number-13: 978-1-4665-6892-1 (Paperback)

Library of Congress Cataloging-in-Publication Data

Huang, Samuel H.
  Supply chain management for engineers / author, Samuel H. Huang.
    pages cm
  Includes bibliographical references and index.
    ISBN 978-1-4665-6892-1 (pbk.)
    1. Production management. 2. Business logistics. 3. Mathematical optimization. I. Title.

TS155.H765 2013
658.5--dc23   2013014248

Visit the Taylor & Francis Web site at
http://www.taylorandfrancis.com

and the CRC Press Web site at
http://www.crcpress.com

Contents

Preface

Supply chain management is an area that has generated a great deal of interest in academia over the past two decades. The emergence of supply chain management is a response to increased competition in the global market. It integrates suppliers, manufacturers, distributors, and customers to reduce costs over the entire chain and to respond quickly to customer needs. Supply chain management is inherently multidisciplinary and requires the efforts of both engineers and business managers. Broadly speaking, engineers deal with modeling and optimization of supply chain operations, whereas business managers focus on strategies for the framework of supply chain models. In practice, business strategy dictates supply chain strategic decisions, whereas modeling and optimization are tools to facilitate the execution of these decisions.

Supply chain management was originally taught in business schools with a strong focus on strategic thinking. It has since appeared as elective and graduate courses in engineering colleges. It is a logical conclusion that more and more engineering colleges will offer this course in order to satisfy increased demand of engineers with supply chain knowledge. However, the majority of supply chain management textbooks are geared toward business school students. Some of these books do not cover quantitative analysis techniques; others include only a superficial coverage of modeling and optimization techniques. Supply chain management textbooks that have a reasonable coverage of quantitative analysis techniques are few and far between. Typically, the presentation of these techniques was intertwined with a large amount of qualitative discussion of strategic issues. For an engineering course focusing on modeling and optimization, one has to distill the quantitative analysis components and supplement them with more in-depth coverage that is suitable for the engineering discipline.

This book is targeted toward engineering students. It is concise, straightforward, and easy to read. It is also strongly influenced by the problem-based learning (PBL) pedagogy. It uses example problems to introduce key concepts. Case studies are then used to help strengthen students' analysis and synthesis skills. It also presents a simulation game where students can play the roles of suppliers, Original Equipment Manufacturers, and retailers within a supply chain environment to practice the skills they acquired.

This book is designed to cultivate students' practical problem-solving skills. It covers both theoretical concepts and the use of practical software tools, including Excel spreadsheet and Gurobi Optimizer. The majority of the examples and all the case problems are presented in a real-world application context. Students are encouraged to act as consultants to provide solutions and recommendations. This provides a motivation for students to acquire supply chain knowledge and to sharpen their problem-solving skills.

It is recommended that the instructor use a combination of traditional lecture and PBL approach when using this textbook. PowerPoint lecture slides with embedded Excel solutions and Gurobi model files are provided. The instructor can start by presenting key supply chain concepts and modeling and optimization techniques using the lecture slides. The students are encouraged to practice problem solving using the embedded Excel solution and Gurobi model files. The instructor can then dedicate a few class sessions on selected case studies where students are required to work together to develop solutions. During these class sessions, the instructor should serve as a facilitator to guide and challenge the learning process. In addition to exercises, this book also provides several problems that are relatively complicated. These problems can be used as mini-projects for the students so they will have an opportunity to link theoretical concepts to practical problem solving. The instructor can select a few students to present their solutions to the entire class. This will provide an opportunity to engage students in peer-to-peer learning. It also makes the students' thinking processes transparent so the instructor can better assess their knowledge, skill, and attitude. Finally, the simulation game can be used as a class project. Most students are motivated to make their company as profitable as possible. They also appreciate the opportunity to engage in contract negotiation where not only analytical skills but also interpersonal skills play an important role. It is a fun experience for the students as well as for the instructor.

MATLAB® is a registered trademark of The MathWorks, Inc. For product information, please contact:

The MathWorks, Inc.
3 Apple Hill Drive
Natick, MA 01760-2098 USA
Tel: 508-647-7000
Fax: 508-647-7001
E-mail: info@mathworks.com
Web: www.mathworks.com

Author

Samuel H. Huang is professor and director of Intelligent Systems Laboratory at the School of Dynamic Systems, the University of Cincinnati, Cincinnati, Ohio. He was previously assistant professor of industrial engineering at the University of Toledo (1998–2001) and systems engineer at EDS/ Unigraphics (1996–1997, now Siemens PLM Software). He received his BS in instrument engineering from Zhejiang University, Hangzhou, People's Republic of China, in 1991 and his MS and PhD in industrial engineering from Texas Tech University, Lubbock, Texas, in 1992 and 1995, respectively. Dr. Huang's research focuses on supply chain management, system analysis and optimization, and predictive analytics, with applications in manufacturing and healthcare delivery. He received the Robert A. Dougherty Outstanding Young Manufacturing Engineer Award from the Society of Manufacturing Engineers (SME) in 2005. Dr. Huang has published over 120 technical papers. He serves on the editorial boards of *International Journal of Industrial and Systems Engineering, Applied Computational Intelligence and Soft Computing*, and *Recent Patents on Computer Science* and on the advisory board of *International Journal of Advanced Manufacturing Technology*. In addition to many industry-sponsored projects, Dr. Huang has been awarded five grants from the National Science Foundation for his research in manufacturing, healthcare analytics, and engineering education.

1

Designing and Engineering the Supply Chain for Competitive Advantage

1.1 Overview

Generally speaking, a *supply chain* is a network of facilities that procure raw materials, transform them into intermediate goods and then final products, and deliver the products to customers through a distribution system. It encompasses all the information, financial, and physical flows from the supplier's supplier to the customer's customer. Consider the automotive industry; the development, design, production, marketing, and delivery of new cars is a team effort that begins with extracting raw materials from the earth, continues through design, fabrication, and assembly, and ends with fit and finish in the dealer's showroom. When a customer buys a car, the customer chooses the output of the entire supply chain and pays all the participants. To be successful, automotive companies must develop an approach to design, organize, and execute supply chain activities from its roots in basic materials extracting to the dealer network. This does not mean ownership or even direct control, but it does imply mechanisms that influence decision making and impact system-wide performance.

A typical supply chain involves the following stages: (1) raw material/component suppliers, (2) original equipment manufacturers (OEMs), (3) wholesalers/distributors, (4) retailers, and (5) customers. Consider the case of a customer purchasing an HTC mobile phone from an electronics store at MBK Bangkok, which is one of the best shopping malls in Thailand. The electronics store, a retailer, offers a variety of mobile phones and other electronics products for the customer to choose from. It obtains mobile phones from Brightstar, a distributor of various wireless devices produced by Apple, Nokia, HTC, and other OEMs. HTC, an OEM, obtains metal cases for its mobile phones from Catcher Technology, a component supplier. The customer, MBK electronics store, Brightstar, HTC, and Catcher Technology are different stages in a mobile phone supply chain. Each stage is connected through the flow of materials, information, and funds. A distinct process

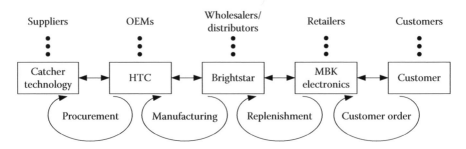

FIGURE 1.1
A typical supply chain and its process cycles.

cycle occurs between two successive stages of the supply chain, resulting in a total of four process cycles, as shown in Figure 1.1. These process cycles are summarized as follows:

- *Customer order process cycle.* This process cycle occurs between the customer and the retailer. A customer arrives at the MBK electronics store and decides to purchase an HTC mobile phone. The store processes this information and fulfills the customer's order. The customer receives the HTC mobile phone that he/she desires. Funds are transferred from the customer to the electronics store.

- *Replenishment process cycle.* This process cycle occurs between the retailer and the wholesaler/distributor. As the inventory of mobile phones at the MBK electronics store decreases to a certain level because of fulfilling customer orders, the electronics store places an order to Brightstar. Brightstar processes this information and sends a shipment to the electronics store. The electronics store replenishes its inventory. Funds are transferred from the electronics store to Brightstar.

- *Manufacturing process cycle.* This process cycle occurs between the wholesaler/distributor and the OEM. To replenish its inventory of HTC mobile phones, Brightstar places an order to HTC. HTC schedules its production based on order information. It then produces and ships the ordered mobile phone to Brightstar. Brightstar replenishes its inventory. Funds are transferred from Brightstar to HTC.

- *Procurement process cycle.* This process cycle occurs between the OEM and the raw material/component supplier. HTC orders metal cases for its mobile phones from Catcher Technology based on its production schedule. Catcher Technology schedules its production based on order information. It then produces and ships the ordered metal cases to HTC. HTC receives the metal cases for its production. Funds are transferred from HTC to Catcher Technology.

Note that the previous description is a simplified version of a mobile phone supply chain focusing on a single entity in each stage of the supply chain. In reality, the MBK electronics store sells products obtained from different wholesalers/distributors. It could also obtain products directly from certain OEMs. Brightstar supplies thousands of retailers worldwide. When it orders mobile phones from HTC, Brightstar aggregates demand from all the retailers it supplies. In addition to HTC, Brightstar provides value-added distribution services for many other OEMs. On the other hand, HTC also provides mobile phones to different wholesalers/distributors. Other than mobile phones, HTC manufactures products such as tablets. When it schedules production, HTC aggregates demands of different products from all of its customers. In the production process, HTC requires many other components in addition to metal cases for mobile phones. Therefore, it has many suppliers in addition to Catcher Technology. Besides HTC, Catcher Technology supplies metal cases to other OEMs as well. Therefore, it also uses aggregated demand information from all of its customers for production scheduling. In summary, the supply chain relationships among different companies are complex and require a systematic management strategy.

A supply chain derives its revenue from the customer. It incurs various costs in providing products and services to the customer. The difference between the revenue generated and the overall costs across the supply chain is the *supply chain profitability*. This is the profit to be shared across all supply chain stages. If each stage of the supply chain aims to maximize its own profit without considering the entire supply chain, it may result in lower supply chain profitability. Therefore, the objective of a supply chain should be to maximize its profitability by carefully managing its assets, products, information flow, and logistics.

The idea of *supply chain management* is to view the chain as an integrated system and to fine-tune the decisions about how to operate the various components (companies, functions, and activities) in ways that can produce the most desirable overall system performance in the long run. Supply chain management is made possible by the dramatic development of information technology in the last several decades, while its seed was sown during World War II when manufacturers were using systems capable of mass producing standardized products efficiently. Emerging from World War II, demand for all kinds of consumer products created large back orders. In this environment, variety was limited and a manufacturer's competitive strength lay in low-cost production and exploiting economies of scale. In the late 1970s, with a decline in their share of the world markets, U.S. manufacturers recognized that the price and quality of their products made them less competitive. They thus adopted the total quality management (TQM) paradigm and regarded price and reduced lead time as the market qualifier. Gradually, customers began to demand high-quality products with low price and short lead time. This forced manufacturers to reexamine the mass production concept to reveal previously "hidden" costs (due to the prevalent costing system at the time) such as inventory, quality, and overheads. Another realization was that success in the global economy would be achievable only if distinct market segments were identified and targeted and products were custom-made to suit the customer's individual needs. Manufacturers were thus forced to develop new methodologies and tools to face product customization and lead time reduction. This gave rise to the supply chain management concept, which integrates suppliers, manufacturers, wholesalers/distributors, retailers, and customers to (1) reduce costs over the entire chain and (2) respond quickly to customer needs.

Supply chain management is inherently multidisciplinary and requires the efforts of both engineers and business managers. Broadly speaking, engineers deal with design and development of technologies and models of supply chain, whereas business managers work on the development and implementation of strategies for the framework of supply chain models. In practice, business strategy dictates supply chain strategic decisions, whereas technologies and models are tools to facilitate the execution of these decisions. Therefore, successful supply chain management requires both the understanding of the strategies behind supply chain design and the mastery of supply chain modeling and optimization techniques.

This book is written for engineers with an emphasis on modeling and optimization techniques. However, engineers also need to have a clear understanding of strategic issues in supply chain management. The rest of the chapter will provide an overview of such issues, including the types of supply chains, methods for designing supply chains, supply chain process models and performance metrics, and drivers to improve supply chain performance.

1.2 Supply Chain Types

Vonderembse et al. (2006), after a careful review of the literature, defined three types of supply chains, summarized as follows:

- *Lean supply chain.* A lean supply chain (LSC) employs continuous improvement efforts that focus on eliminating waste or nonvalue-added steps along the chain. It is supported by efforts to achieve internal manufacturing efficiencies and setup time reduction, which enable the economic production of small quantities and enhance cost reduction, profitability, and manufacturing flexibility to some degree. The short setup time provides internal flexibility, but an LSC may lack external responsiveness to customer demands, which can require flexibility in product design, planning and scheduling, and distribution in addition to manufacturing.

- *Agile supply chain.* An agile supply chain (ASC) focuses on responding to unpredictable market changes and capitalizing on them through fast delivery and lead time flexibility. It deploys new technologies, methods, tools, and techniques to solve unexpected problems. It utilizes information systems and technologies, as well as electronic data interchange capabilities to move information faster and make better decisions. It places more emphasis on organizational issues and people (knowledge systems and empowered employees), so decision making can be pushed down the organization. The ASC paradigm is a systemic approach that integrates the business, enhances innovations across the company, and forms virtual organizations (VOs) and production entities based on customer needs.

- *Hybrid supply chain.* A hybrid supply chain (HSC) generally involves "assemble to order" products whose demand can be forecasted with a relative high level of accuracy. The chain helps to achieve mass customization by postponing product differentiation until final assembly. The lean or agile supply chain techniques are utilized for component production with different characteristics. For example, in the automotive industry, air bags would most likely be produced with an LSC while engine electronics might require the innovation found in an ASC. In addition, the agility is needed to establish a company–market interface to understand and satisfy customer requirements by being responsive, adaptable, and innovative.

A more detailed description of LSC, ASC, and HSC is provided in Table 1.1. These supply chain types are, to a large extent, a function of product characteristics and customer expectations (Fisher 1997). With a rapidly changing business environment, organizations require a supply chain model that deals with strategic and customer issues in addition to operating constraints. The product is the soul of the supply chain; there is no justification for adopting a particular supply chain type unless it conforms to the needs of the product and its customers. Three case studies are presented next to help the reader better understand the relationships between supply chain types and product characteristics.

Case Study 1.2.1 Black and Decker's Lean Supply Chain

Black and Decker Inc. (now Stanley Black & Decker) produces a variety of small appliances and hand tools for use in the home. Success in that market is predicated on manufacturing products that have high quality and low cost and a moderate amount of variety. Designs for these appliances and tools change slowly, and demand for these products can be characterized as slow and steady growth. Under these circumstances, an LSC effectively meets the needs of Black and Decker. An LSC focuses on operating issues as it attempts to eliminate nonvalue-added operations.

TABLE 1.1

Summary of Lean, Agile, and Hybrid Supply Chains

Category	Lean Supply Chain (LSC)	Agile Supply Chain (ASC)	Hybrid Supply Chain (HSC)
Definition	An LSC employs continuous improvement to focus on the elimination of waste or nonvalue-added steps in the supply chain. It is supported by the reduction of setup time to allow for the economic production of small quantities, thereby achieving cost reduction, flexibility, and internal responsiveness. It does not have the ability to mass customize and be adaptable easily to future market requirements	Agility relates to the interface between a company and the market. ASCs profit by responding to rapidly changing, continually fragmenting global markets by being dynamic and context-specific, aggressively changing, and growth oriented. They are driven by customer-designed products and services	A HSC generally involves "assemble to order" products, where demand can be accurately forecasted. The supply chain helps to achieve some degree of customization by postponing product differentiation until final assembly. Lean or agile supply chains are utilized for component production. The agile part of the chain establishes an interface to understand and satisfy customer requirements by being responsive and innovative
Purpose	Focus on cost reduction and flexibility for already available products. Employs a continuous improvement process to focus on the elimination of waste or nonvalue–added activities across the chain. Primarily aims at cost cutting, flexibility, and incremental improvements in products	Understands customer requirements by interfacing with customers and market and being adaptable to future changes. Aims to produce in any volume and deliver to a wide variety of market niches simultaneously. Provides customized products at short lead times (responsiveness) by reducing the cost of variation	Employ lean production methods for manufacturing. Interfaces with the market to understand customer requirements. Achieve a degree of customization by postponing product differentiation until final assembly and adding innovative components to the existing products
Approach to manufacturing	Advocates lean manufacturing techniques	Advocates agile manufacturing techniques, which is an extension of lean manufacturing	Employs lean and agile manufacturing techniques

TABLE 1.1 (continued)

Summary of Lean, Agile, and Hybrid Supply Chains

Category	Lean Supply Chain (LSC)	Agile Supply Chain (ASC)	Hybrid Supply Chain (HSC)
Integration	Integrate manufacturing, purchasing, quality, and suppliers	Integrates marketing, engineering, distribution, and information systems	Similar to the LSC at component level and follows an ASC at product level
Production planning	Works on confirmed orders and reliable forecasts	Has the ability to respond quickly to varying customer needs (mass customization)	Works on confirmed orders and reliable forecasts with some ability to achieve some produce variety
Length of product life cycle	Standard products have relatively long life cycle time (>2 years)	Innovative products have short life cycle time (3 months to 1 year)	Involved the production of "assemble to order" products, which stay in the maturity phase of the life cycle for a long time
Alliances	May participate in traditional alliances such as partnerships and joint ventures at the operating level	Exploits a dynamic type of alliance known as a "virtual organization" that works on product design	Along with traditional operating alliances, HSCs may utilize strategic alliances to respond to changing consumer requirements
Markets	Serve only the current market segments	Acquire new competencies, develop new product lines, and open up new markets	Respond to customer requirements with innovative features in existing products. This enables the organization to capture a larger segment of that product market
Organizational structure	Uses a static organizational structure with few levels in the hierarchy	Create virtual organizations by creating alliances with partners that vary with different product offerings that change frequently	Maintain an organization similar to an LSC. May create temporal relationships with partners to implement innovative features
Approach to choosing suppliers	Supplier attributes involve low cost and high quality	Supplier attributes involve speed, flexibility, and quality	Supplier attributes involve low cost and high quality, along with the capability for speed and flexibility, as and when required

(*continued*)

TABLE 1.1 (continued)

Summary of Lean, Agile, and Hybrid Supply Chains

Category	Lean Supply Chain (LSC)	Agile Supply Chain (ASC)	Hybrid Supply Chain (HSC)
Demand patterns	Demand can be accurately forecasted and average margin of forecasting error tends to be low, roughly 10%	Demand is unpredictable with forecasting errors exceeding 50%	Similar to the LSC. The average product demand can be accurately forecasted. Component-level forecasting may involve larger errors
Inventory strategy	Generates high turns and minimizes inventory throughout the chain	Make in response to customer demand	Postpone product differentiation and minimize functional components inventory
Lead time focus	Shorten lead time as long as it does not increase cost	Invest aggressively in ways to reduce lead time	Similar to the LSC at component level (shorten lead time but not at the expense of cost). At product level, to accommodate customer requirements
Manufacturing focus	Maintain high average utilization rate	Deploy excess buffer capacity to ensure that raw material/ components are available to manufacture the innovative products according to market requirements	Combination of lean and ASC depending on the company
Product design strategy	Maximize performance and minimize cost	Design products to meet individual customer needs	Use modular design in order to postpone product differentiation for as long as possible
Human resources	Empowered individuals working in teams in their functional departments	Involves decentralized decision making. Empowered individuals working in cross-functional teams, which may be across company borders too	Empowered individuals working in teams in their functional departments

Source: Modified from *Int. J. Prod. Econ.,* 100 (2), Vonderembse, M.A., Uppal, M., Huang, S.H., and Dismukes, J.P., Designing supply chains: Towards theory development, 223–238, Copyright 2006, with permission from Elsevier.

LSC partners support the reduction of setup time to enable the economic production of small quantities. This enables the supply chain to keep inventory costs low, achieve manufacturing cost reductions, and enables manufacturing operations to switch quickly among products, which provides a degree of responsiveness to customer needs. Customization of individual products to satisfy specific requirements is not necessary because a standard product meets the needs of most customers at an affordable price.

As an example, consider the 3/8-in., variable speed, reversing drill, which is one of Black and Decker's most popular products. The tool is sold primarily to homeowners, who use it infrequently to hang a shelf or repair a table. The major components of the drill, whether produced by internal or external suppliers, can be divided into three major groups. The electric motor and other electronic components provide the power, the gearing transforms the power to the drill bit, and the housing supports and encases the motor and the gears. There are other items, including fasteners, that are also purchased, but these items are standard and can be provided by many suppliers.

To create a successful supply chain, component suppliers must adopt lean manufacturing and its continuous improvement philosophy. These suppliers must achieve an interesting combination of flexibility and cost reduction. Flexibility is needed because there are several different models of drills, as well as other hand tools and appliances that require similar components. Cost reduction is also essential because products, like drills, are produced by many competitors and customers are price sensitive. Cost reductions can be achieved when suppliers purchase large volumes of basic materials such as steel for the gear manufacturer or copper for the electric motor producer. They are also achieved by streamlining the flow of materials and information through the supply chain to drive out inventory and nonvalue-added steps. Because drills have low profit margins, maintaining high sales and production volumes is critical for profitability for all members in the supply chain.

Changes in product design are incremental and often focus on small improvements in performance or cost reductions. Substantial improvements in product performance are not generally available. For example, most drills today have a keyless chuck so that bits can be quickly replaced without searching for a tool to remove the bit. This redesign was accomplished easily and quickly. In another case, parts of the housing are now made of plastic rather than steel because plastic is lighter and cheaper. As a result, Black and Decker can with relative ease switch from one supplier of electric motors to another, which is a significant motivator for suppliers to seek continuous improvements in both component part cost and quality. To the extent that Black and Decker and its supplier achieve this, they will maintain and expand their share of the market.

Case Study 1.2.2 IBM's Agile Supply Chain

IBM operates in the highly competitive information technology arena, where response to rapidly changing and continually fragmented global markets is essential. This requires IBM to move beyond operating partnerships that cut manufacturing costs and reduce manufacturing time to an environment that creates strategic partnerships that work jointly on research and development, product conceptualization, product development, and distribution, as well as operations. Under these circumstances, an ASC creates dynamic and context-specific partnerships with various companies in order to meet specific customer needs. An ASC learns and assimilates requirements by interfacing with customers and markets. It adapts to changing expectations quickly and with minimal disruption. It provides customized products at short lead time by reducing the cost of variation.

IBM pursued a strategic business alliance with Hitachi that is designed to accelerate the development and delivery of advanced storage technologies and products to meet diverse customer expectations. This alliance combines technical leadership with global economies of scale in designing and manufacturing disk storage. Various hard disk drive operations were reorganized into a new standalone joint venture that integrated research, development, and manufacturing. It also allowed the organizations to coordinate related sales and marketing teams. In the storage area, IBM has also collaborated with Tree Data to cater to individual customer requirements in a timely manner. This collaboration helps IBM concentrate on storage networking products for mid-sized customers. With the flexibility that this partnership brings, IBM can design and develop storage systems that meet the needs of various market segments easily and quickly.

IBM is collaborating with UPS to manage IBM's distribution network, which covers Asia, Europe, and North America. The partnership was created because UPS has the skills and ability to move products easily, quickly, and efficiently. UPS can help IBM coordinate and integrate the supply and distribution networks that link suppliers to manufacturers and manufacturers to customers. These efforts expand collaboration among the four logistic centers in Singapore, Taiwan, the Netherlands, and the United States, and 22 just-in-time suppliers and other vendor-managed inventory locations. This has increased the real-time visibility throughout the supply chain for IBM, which improves the management of inventory and shortens product turnaround time. In addition, UPS holds licenses in Europe and Asia that allow "self-reporting" of duties and taxes owed after shipment, which reduces paperwork and creates seamless operations in the supply chain. The important benefits achieved through this collaboration are shorter cycle time from manufacturers to customers, doubling inventory turns, and flexibility in expedited service and rush deliveries.

Case Study 1.2.3 DaimlerChrysler's Hybrid Supply Chain

Because vehicles are complex, with hundreds or even thousands of components of varying cost and sophistication, DaimlerChrysler (now Chrysler Group LLC) cannot use a one-size-fits-all approach when developing supplier relationships. Some components that have high product and/or process technology are undergoing rapid technological change and can add significant value to the vehicle in the eyes of the consumer. Global positioning and information systems for navigation are examples. This type of component may require DaimlerChrysler to use ASC and to develop a strategic partnership with the supplier in order to work jointly on product design and development. For other components, product technology is well established but the components themselves are high cost, bulky, and subject to variation. These products also contribute significantly to factors that impact the customer decision to purchase or repurchase such as comfort, safety, appearance, or reliability. For example, vehicle seats consume a lot of space, and they come in a variety of styles and colors. Without careful management and coordination between suppliers and manufacturers, production, inventory, and material handling costs can spiral out of control. In this case, DaimlerChrysler attempts to create operating partnerships that improve overall supply chain performance. Some components are basic commodities that customers do not see or appreciate, but they are essential to the vehicle's performance. These hidden items, like hose couplings and wire connectors, are not points of differentiation for the customer, and they are not valued by customers in the traditional sense. Customers only notice them when they fail. These items are purchased based on quality, cost, and reliable and on-time delivery. This wide range of components presents DaimlerChrysler with fundamentally different supply chain management issues. The following are three examples that illustrate how the company handles these alternatives:

- *Strategic Partnership with Dana Corporation.* Dana designs and manufactures drive train components for cars and trucks, which are key components for delivering engine performance to the wheels, improving operating efficiency, and enhancing comfort. Many vehicles, especially light trucks and sport utility vehicles, have more than one gearing option, so platform design and flexibility in manufacturing are also important. Drive train components interact significantly with engine and body design, and engineers are beginning to use computer chip to improve their efficiency. To optimize vehicle performance, DaimlerChrysler, Dana Corporation, and other first-tier suppliers work together on design teams to address important design issues and make critical tradeoffs. Changes in weight, engine displacement, or traction requirements can impact design decisions for Dana. In addition, Dana's manufacturing operations are designed to achieve high quality, keep costs low, and cope with variability in demand for the components. DaimlerChrysler defines metrics to assess supplier performance, sets target levels for these metrics, measures outcomes, and works with suppliers to improve performance.

- *Operating Partnership with Modine Manufacturing.* Modine Manufacturing supplies cooling modules to DaimlerChrysler's new facility that assembles the Jeep Liberty. A cooling module includes a radiator, fan, condenser for the air conditioner, oil cooler, wiring, supporting frame, and hose connections. The modules are assembled by Modine and delivered to DaimlerChrysler's final assembly plant ready for installation. To keep inventory and material handling costs low, DaimlerChrysler insists that Modine deliver the cooling modules within 4 h of the order and in the sequence needed at the assembly line. To do this, DaimlerChrysler sends an electronic signal to the Modine Plant every time a vehicle begins the trip down the assembly line (every 72 s). When Modine gets the signal, it assembles the module from materials fabricated at other facilities or it pulls the module from a modest 2-day inventory. The modules are loaded on shipping racks in the order needed at final assembly and delivered within 4 h. Upon arrival, the rack is taken directly to the assembly line where the modules are taken from the rack and placed in the vehicle. This is much better than the traditional approach, which would require DaimlerChrysler to stage the product, check the order, determine what was needed immediately, and place what was not needed in inventory. Modules placed in inventory would be picked from inventory eventually and taken to the assembly line. DaimlerChrysler pays only for modules that are in vehicles that drive off the assembly line. This new approach eliminates unneeded paperwork and clerical activity, and it greatly reduces inventory and material handling costs. To help Modine and its suppliers determine their inventory needs, DaimlerChrysler provides a 5-day rolling schedule for production that is not fixed but is usually very accurate.

- *Commodity Purchasing: Threaded Fasteners.* DaimlerChrysler also purchases many different threaded fasteners for the final assembly process. The size, shape, and strengths of these fasteners are highly standardized. There are few, if any, significant design decisions. These fasteners are simple and take little storage space, so close coordination of production, shipping, and delivery of the fasteners is of limited importance. The primary elements in the purchasing decision are quality, price, and reliable delivery. Strategic and operating partnerships add little value. E-purchasing practices and competitive bidding can be used to streamline paperwork and coordinate delivery. The Internet has become an effective means for purchasing these commodities because it eliminates transactions that increase cost but add little if any value.

1.3 Supply Chain Design

The goal of a supply chain is to meet specific customer expectations and in the mean time generate profit for the entire chain. As previously mentioned, supply chain design should be a function of the product characteristics and

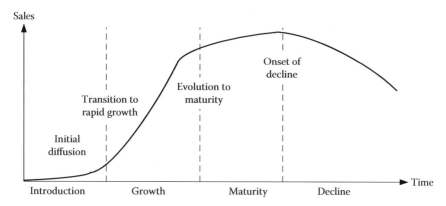

FIGURE 1.2
Product life cycle sales curve.

expectations of the final customer. Products usually go through a four-stage life cycle—introduction, growth, maturity, and decline (Day 1981). The shape of the sales curve (Figure 1.2) reflects the notion that a product's sales begin slowly during the introductory stage, then grow rapidly, often reaching a peak relatively early in a product's life. During the saturation or maturity phase, demand may grow slowly for a long period of time before it begins to decrease. Accompanying this change in real growth are changes in competitive conditions, strategies, and performance. The life cycle can be viewed as a series of interrelated propositions dealing with systematic changes in the marketplace.

Different products have different demand patterns and other characteristics. They can be categorized into three types, summarized as follows:

- *Functional products.* Functional products have stable demand, and their design characteristics and production requirements change slowly over time. Commodities like staples or fasteners are functional products that require straightforward supply chains with few participants. More interesting examples of functional products would be small appliances or hand tools like toasters or saber saws because they tend to have several suppliers providing important components. These products are usually in the latter part of the growth segment of their product life cycle or beyond.

- *Innovative products.* Innovative products are new or derivative products that are aimed at new customers and markets and are designed to be adaptable to changing customer requirements. These products require close and continuous customer contact, have uncertain demand, and their product designs may be unstable (Fisher 1997, Mason-Jones et al. 2000). Innovative products are usually in the introduction and growth stages of the product life cycle.

Emerging communication technology such as smart phones is an example. Innovative products can also be derivative or differentiated products that reignite the growth potential of a product in the mature phase of the product life cycle. New computer chips and software upgrades are examples.

- *Hybrid products.* Hybrid products are complex products that have a variety of components, which may be a mixture of functional and innovative products. Automobile or other assembled products are examples. These products are usually major purchases that are made periodically by customers after careful consideration and investigation.

Huang et al. (2002) proposed a framework of matching product types to supply chain types. A more formal theoretical development was later provided by Vonderembse et al. (2006). The framework is summarized in Table 1.2 and described as follows.

An LSC focuses on reducing lead time, increasing efficiency, expanding manufacturing flexibility, and cutting cost. It attempts to build a level schedule across the supply chain, and it responds to production pulled by customer demand. While striving for these goals, the LSC focuses on incremental improvements (kaizen). It tries to improve the product and the associated processes by balancing the supply chain. The long product life cycle of functional product provides a stable design over many years. Being a low-cost item with highly predictable demand patterns, profitability can be achieved by minimizing cost and employing a level schedule over the entire supply chain and over all the product's life cycle stages. This justifies the usage of an LSC for functional products. In addition to providing cost minimization, the LSC is efficient and flexible, and it brings about incremental improvements, permitting companies to constantly improve the quality of their products to keep their customers satisfied. In summary, to have the

TABLE 1.2

Matching Product Types to Supply Chain Types

	Product Type		
Product Life Cycle	**Functional**	**Innovative**	**Hybrid**
Introduction		Agile supply chain	
Growth	Lean supply chain		Hybrid supply chain
Maturity		Lean supply chain	
Decline			

Source: Modified from *Int. J. Prod. Econ.*, 100 (2), Vonderembse, M.A., Uppal, M., Huang, S.H., and Dismukes, J.P., Designing supply chains: Towards theory development, 223–238, Copyright 2006, with permission from Elsevier.

highest internal performance and customer satisfaction, functional products should be designed and produced by LSCs through all stages of the product life cycle.

In order for innovative products to succeed, they should be producible in any volume, as per customer requirements. The first two stages of the product life cycle, introduction and growth, are the testing grounds to ensure that organizations are achieving customization and market adaptability. For this, one of the strategic tools provided by ASC is a VO. A VO integrates complementary resources existing in a number of cooperating companies to produce a particular innovative product as long as it is economically justifiable to do so. This dynamic alliance provides access to a wide range of world class competences, enabling organizations to overcome the customization/responsiveness squeeze. This supports the usage of an ASC for the first two stages in the life cycle of an innovative product. By providing concurrency of operations among the members of the VO, the agile company can rapidly deliver its innovative products in small quantities, as per customer requirements. In summary, to have the highest internal performance and customer satisfaction, innovative products should be designed and produced by ASCs in the introduction and growth stages of the product life cycle.

Once the product has been firmly established, it transitions into the third stage of its life cycle, maturity. By this stage, the innovative product begins to take on the characteristics of a functional product. Price competitiveness becomes very important. Production becomes more routine and is part of the organization's daily schedule, which has already followed the LSC concept. In order to maximize their profits, organizations still need to deal with their customers and not only provide them with the support they need but also introduce new, improved versions of the existing product, thus maintaining their customer base. From the maturity level onward, an organization may employ LSC to meet the needs of this product. Therefore, to have the highest internal performance and customer satisfaction, innovative products should be designed and produced by LSCs in the maturity and decline stages of the product life cycle.

Hybrid products, which are complex, require the organization to bring together a set of suppliers with a wide range of capabilities. This implies innovative and functional components, as well as strategic partnerships. While it may be true that hybrid products that are near the end of their product life cycle may use fewer innovative components than a product that is at the beginning of its life cycle, there is always the opportunity to introduce innovation into hybrid products. As a result, hybrid products require HSC throughout their product life cycles. In other words, to have the highest internal performance and customer satisfaction, hybrid products should be designed and produced by HSCs throughout the product life cycle.

1.4 Supply Chain Process Model

The supply chain council (http://supply-chain.org), a not-for-profit organization established in 1996, has developed the supply chain operations reference (SCOR) model to support supply chain strategic decision making. Since then, the SCOR model has been used by many companies throughout the world. Some examples are summarized as follows (Huang et al. 2005, Zhou et al. 2011):

- Intel started implementing the SCOR model in its Resellers Product Division in 1999 and later expanded the implementation to its Systems Manufacturing Division. The benefits include shorter cycle time, reduced inventory, improved supply chain visibility, and timely access to important customer information.
- General Electric's Transportation Systems unit applied the SCOR model to streamline the purchasing process with its suppliers. The result is shorter purchasing cycle time and lower cost.
- Phillips Lighting used the SCOR model in its overall business framework to improve customer service and reduce inventory.
- A New York hospital used the SCOR model to improve its supply chain performance. It reported 8%–10% reduction in excess and obsolete inventory, 21% increase in capacity, 8% increase in demand, and up to 40% reduction in prep times for key procedures.
- Avon used the SCOR model to simplify its supply chain process, optimize the manufacturing location of different products, centralize inventory management, and rationalize supplier base. After SCOR implementation, perfect order rate increased from 62% to 90% and cycle time is reduced by 50%.
- LEGO utilized SCOR best practice to support product allocation, performance measurement, information technology integration, vendor managed inventory, and automated order processing. As a result, its capacity increased by 10%, inventory days of supply increased from 55 to 75 days, and delivery performance increased from 70% to over 90%.
- Siemens Medicals used the SCOR model to support the transfer of its supply chain processes to an e-enabled world. Specifically, SCOR standards were used to develop solutions that can be reused as much as possible within the company. Delivery lead time was reduced to 2 weeks from 22 weeks, and delivery reliability was increased to 99.5% from 65%.

Some researchers have provided review and analysis of the SCOR model (Huang et al. 2004, 2005). The model is continuously evolving with direct inputs from industry practitioners who manage global supply chains and use it to analyze and improve the performance of their organizations. A brief discussion of the SCOR model is provided here based on the materials presented in Huang et al. (2005).

The SCOR model is a process reference model. Specifically, it is a model that links process elements, metrics, best practice, and the features associated with the execution of a supply chain in a unique format. It is designed to be configurable and aggregates a series of hierarchical process models. The use of a process reference model allows companies to communicate using common terminology and standard descriptions of the process elements that help understand the overall supply chain management process and the best practices that yield the optimal overall performance.

The SCOR model is a model that can be used to configure the supply chain based on business strategy. It provides unambiguous, standard descriptions for various activities within the supply chain. It also identifies the performance measurements and supporting tools suitable for each activity. This process reference model enables all departments and businesses involved in developing and managing the integrated supply chain to collaborate effectively.

The SCOR model integrates the well-known concepts of business process reengineering, benchmarking, and process measurement into a cross-functional framework. It captures the "as-is" state of a process and then derives the desired "to-be" future state. It quantifies the operational performance of similar companies and establishes internal targets based on "best-in-class" results. It also characterizes the management practices and software solutions that result in "best-in-class" performance. The structural framework of the SCOR model is composed of the following elements:

- Standard descriptions of the individual elements that make up the supply chain processes
- Standard definitions of key performance measures
- Descriptions of best practices associated with each of the process elements
- Identification of software functionality that enables best practices

The SCOR model has five distinct management processes, namely, *Plan*, *Source*, *Make*, *Deliver*, and *Return*, which are called Level 1 processes. The Plan process consists of processes that balance aggregated demand and supply to develop a course of action that best meets the business goals. Plan processes deal with demand/supply planning, which include the activities to assess supply resources, aggregate and prioritize demand requirements, and plan inventory, distribution, production, material, and rough-cut capacity for all products and all channels.

The Source process contains processes that procure goods and services to meet planned or actual demand. Sourcing/material acquisition includes the jobs of obtaining, receiving, inspecting, holding, and issuing material.

Management of sourcing infrastructure includes vendor certification and feedback, sourcing quality, in-bound freight, component engineering, vendor contracts, and vendor payments.

The Make process includes functions that transform goods to a finished state to meet planned or actual demand. Make is the core process of the system in which actual production execution takes place. It includes the jobs of requesting and receiving material, manufacturing and testing product, packaging, holding, and/or releasing the product eventually.

The Deliver process consists of processes that provide finished goods and services to meet planned or actual demand. This typically includes the functions of order management, transportation management, and distribution management. Managing the deliver process includes managing channel business rules, ordering rules, managing deliver inventories, and managing deliver quality.

The Return process deals with the reverse flow of material and information related to defective and surplus products. This includes authorizing, scheduling, receiving, verifying, disposing and replacement or credit for the aforementioned types of materials. Each basic supply chain is a "chain" of Source, Make, Deliver, and Return execution process. Each interaction of two execution processes is a "link" in the supply chain. Planning sits on top of these links and manages them.

The SCOR model Level 1 metrics characterize performance from customer-facing and internal-facing perspectives. Therefore, at Level 1, basis of competition is defined and broad guidelines are provided to meet the competition. Specific tasks to be completed at Level 1 are set business requirements and define basis of competition, evaluate the performance of current operation vis-à-vis required performance, set the SCOR model metrics and targets, and define the gap, set business priorities, and state what needs to change. Also at Level 1, the current supply chain is modeled considering asset, product volume and mix, and technology requirements and constraints.

Level 2 defines different categories within the Level 1 processes. At this level, processes are configured in line with supply chain strategy. The goal at Level 2 is to simplify the supply chain and enhance its overall flexibility. At Level 2, the SCOR model provides a tool kit of 22 process categories (version 5.0). Any supply chain configuration can be represented with this tool kit. Here, the company should reconfigure the supply chain configured in Level 1 to determine the expected performance. At Level 2, market constraints, product constraints, and company constraints are considered to configure the inter-company and intra-company process categories.

Level 3 allows businesses to define in detail the processes identified, as well as performance metrics and best practices for each activity. The software functionality required to support best practices is also identified, as well as the commercial software tools currently providing the required functionality. Inter-company and intra-company process elements are also defined. Performance levels and practices are defined for these process elements.

Benchmarks and the required attributes for the enabling software are also noted at this level. Specific tasks to be performed at this level include the following: develop process models that support strategic objectives and work within the new supply chain configuration developed at Level 2, set process metrics and performance targets, establish business practices at operating level, build system requirements that support the supply chain configuration, processes, and practices, and finally select appropriate systems. At Level 3, inputs, outputs, and basic logic flow of process elements are captured.

Level 4 describes the detailed tasks within each of the Level 3 activities. These tasks, and their interactions, are unique to each business. This level of detail is needed to implement and manage the supply chain on a day-to-day basis. Level 4 process definition equates to quality process definition (e.g., ISO 9000) in most companies. At Level 4, implementation of supply chain processes takes place. Immediate goals are set, intra-company and inter-company supply chain improvements take place, priorities are set, and rapid results are expected and studied.

The SCOR processes are assigned to three types: planning, execution, and enable. Planning processes plan the whole chain along with planning specific type of execution process. Execution processes cover all process categories of Source, Make, Deliver, and Return, except the enable process categories. Enable process of a particular process type defines the constitution of that particular process element. Using the four levels of the SCOR model, a business can quickly and unambiguously describe its supply chain. A supply chain that is defined using this approach can also be modified and reconfigured rapidly as business and market requirements change. The SCOR model has a powerful role in implementing supply chains. The SCOR model Level 1 and Level 2 metrics keep management focused, while Level 3 metrics support ongoing diagnosis.

Operating a supply chain is far different from running a stand-alone company, and so are the metrics. The supply chain can be viewed as an externalization of business processes toward greater profitability. Trading partners, to a greater or lesser degree of formality, are linking their productive assets to gain efficiencies in cycle times, procurement, inventory, logistics, and cash flow. Given these relationships, how the partners measure effectiveness of their intertwined processes becomes quite different from assessing internal operations. Their shared metrics needed to achieve this balance, which calls for accepted common standards in the field, as well as issues of change management and company culture for all partners. The performance metrics used by the SCOR model have increased steadily as the model evolves. In SCOR model version 10, these metrics are classified into five categories—supply chain reliability, supply chain responsiveness, supply chain agility, supply chain cost, and supply chain management (SCC 2012). A company cannot be the best in all of the metrics. It should wisely target its strength by selecting a number of metrics through which it can differentiate itself in the market, while ensuring that it stays competitive in other metrics.

According to SCC (2012), the organizational benefits of adopting the SCOR model include the following:

- Rapid assessment of supply chain performance
- Clear identification of supply chain performance gaps
- Efficient supply chain network redesign and optimization
- Enhanced operational control from standard core processes
- Streamlined management reporting and organizational structure
- Alignment of supply chain team skills with strategic objectives
- A detailed game plan for launching new businesses and products
- Systematic supply chain mergers that capture projected savings

Zhou et al. (2011) conducted an empirical study to examine the relationships among the supply chain processes as suggested by the SCOR model. The study used a survey research method and the analysis was based on 125 useable responses from U.S. manufacturing companies. The results generally supported the relationships among the supply chain processes in the SCOR model. Specifically, the Plan process was found to have significant positive influence on Source, Make, and Deliver processes. The Source process has significant positive influence on the Make process, whereas the Make process has significant positive influence on the Delivery process. The strongest link found was from the Source process to the Make process, whereas the weakest link found was from the Plan process to the Make process. In addition, the Source process was found to mediate the impact of the Plan process on the Make process, and the Make process was found to mediate the impact of the Plan process on the Delivery process. This study empirically validated the SCOR model structure and gave more credibility to the use of the SCOR model in real-world supply chain management applications. It also showed that the Plan process has received the least attention from the surveyed companies, although it has significant influence on all the other three processes.

1.5 Supply Chain Drivers

Chopra and Meindl (2007) identified three logistical drivers—facilities, inventory, and transportation—and three cross-functional drivers—information, sourcing, and pricing—that determine the performance of a supply chain. These drivers are briefly summarized as follows:

- *Facilities*. Facilities are physical locations in a supply chain where products are manufactured and stored. These include manufacturing plants and distribution centers/warehouses. For a manufacturing plant, a company must decide whether it will be a dedicated facility (to produce a limited number of products efficiently) or a flexible facility (that is capable of producing a variety of products). For a distribution center/warehouse, a company must decide whether it will be primarily a storage facility or a cross-docking facility where inbound products are unloaded, broken into smaller lots, and quickly shipped to customers. The capacities and locations of the facilities must be determined considering the tradeoff of efficiency and responsiveness. A central facility with large capacity is likely to be efficient but not responsive. On the other hand, a number of facilities close to the customers will be responsive but would incur higher costs.

- *Inventory.* Inventory includes all raw materials, work in process, and finished goods in a supply chain. Inventory exists because of a mismatch between supply and demand. It has three components: (1) cycle inventory, which is the average amount of inventory used to satisfy demand between receipts of supplies; (2) safety inventory, which is inventory held in case demand exceeds expectation; and (3) seasonal inventory, which is inventory intentionally built up in low-demand periods for use in high-demand periods. Increasing the level of inventory generally increases supply chain responsiveness (products are likely to be available when customer demand arrives). A high level of inventory also reduces production and transportation costs due to economies of scale. However, the disadvantage is an increase in inventory holding costs.

- *Transportation.* Transportation is the means of moving inventory from point to point in a supply chain. It can take the form of many combinations of modes and routes, including electronic delivery of software products. Faster transportation modes, for example, air, allow a supply chain to be more responsive with increased transportation costs. However, it can reduce inventory holding costs because of the reduction in the required safety inventory. Transportation decisions are usually made in conjunction with facility roles and locations in the context of distribution network design. Operational aspects of the decisions include routing and scheduling, usually with time window constraints (Solomon 1987).

- *Information.* Information consists of data and analysis results concerning the operation of the entire supply chain. It directly affects each of the other drivers and thus is critical to the performance of the supply chain. Information should be shared across the supply chain to achieve supply chain coordination and to maximize supply chain profitability. Information sharing allows a company to better forecast future demand. Forecasting is used both to schedule production and to determine whether to increase production capacity by building new manufacturing plants. Once a forecast is created, a company uses aggregate planning to generate a production plan that can satisfy demand with minimum costs. The aggregate plan should be shared across the supply chain because it contains information about the demand from the company's suppliers and the supply to its customers.

- *Sourcing.* Sourcing is the choice of who will perform a particular supply chain activity. A company must first decide which tasks will be performed in-house and which will be outsourced. For each task to be outsourced, a company needs to select appropriate suppliers and negotiate contracts with them. The procurement process should be structured to improve efficiency and coordination. A company should make sure that a third party can improve supply chain profitability more than the company can before outsourcing a supply chain task.

- *Pricing.* Pricing determines how much a company will charge for its products or services. It affects the behavior of the customers, thus affecting supply chain performance. All pricing decisions should be made with the objective of improving supply chain profitability. Most supply chain activities display economies of scale. Therefore, a common practice is to offer quantity discount. Another pricing strategy is everyday low price. This strategy results in relatively stable demand, which makes it easier to manage inventory. To attract customers with varying needs, differential pricing is used in which a product has different prices based on the type of customer, delivery methods, and payment terms.

Chapter 1 provides a strategic view of the supply chain and how design and engineering tools can be used to create and sustain supply chain competitive advantage. The rest of the chapters focus on specific modeling and optimization techniques used to manage the supply chain. Because a supply chain generates revenue by satisfying customer demands, we start with methods used to forecast demand in Chapter 2. We then proceed to Chapter 3 to present aggregate planning methods, which aim to convert demand forecasts into production plans with minimum costs. Inventory management is the key to satisfy customer demand, which is an important factor that impacts supply chain profitability. Therefore, we dedicate Chapter 4 to techniques used to minimize inventory costs and maximize expected profit. Chapter 5 introduces modeling and optimization techniques used in designing a supply chain distribution network for efficient product delivery, which takes facility, inventory, and transportation into consideration. Chapter 6 discusses techniques for supplier selection, a key issue in supply chain sourcing. In Chapter 7, we present a simulation game where students can play the roles of suppliers, OEMs, and retailers within a supply chain environment to practice the supply chain management skills they acquired.

Problem: Dr. Smart's Supply Chain Strategy

Dr. Smart at the University of Iowa (Iowa City, Iowa) invented a small device (with a high value-to-weight ratio) that can perform house cleaning chores. The main components of this device include a regular computer chip (that are currently used in all PCs), a set of specialized robotic arms, and a sophisticated computer program. Market survey indicates that people from big cities such as New York, Chicago, and Los Angeles are most likely to buy this device. However, the demand for this device is highly uncertain. Dr. Smart wants to establish a company to produce the device. Help him on the following supply chain-related problems:

- There are a lot of companies that can supply computer chips; each can ensure a short lead time if the order size is not too big. Intel can ensure short lead time no matter how large the order size is. However, it requires a long-term contract that prohibits Dr. Smart from using other suppliers (who may offer lower price in the near future). The quality and performance of computer chips from different companies are comparable. Should Dr. Smart sign the contract with Intel? Why?

- ABB and Flexbot are two potential suppliers for the robotic arms. Flexbot has a slightly higher price but is more flexible in terms of scaling down or ramping up production. ABB has a lower price and it also provides quantity discount, but it requires a minimum purchase quantity per month and a long lead time if the order size is large. Which company should be chosen and why?

- The assembly plant for the house cleaning device will be based in the United States. A development center to maintain and enhance the software program is also needed. Should Dr. Smith set up the development center in India (which has talented programmers with a lower cost) or in the United States near the assembly plant (so that programmers can work closely with plant engineers)? Why?

- Dr. Smart is not sure whether he (1) should have separate distribution centers near big cities or (2) should have a central distribution center at Iowa City. What are the pros and cons of these two choices (assuming facility costs are roughly the same)? Which choice makes more sense?

References

Chopra, S. and P. Meindl. 2007. Supply Chain Management: Strategy, Planning, & Operation, 3rd edn. Upper Saddle River, NJ: Pearson Prentice Hall.

Day, G. S. 1981. The product life cycle: Analysis and applications issues. *The Journal of Marketing* **45**: 60–67.

Fisher, M. L. 1997. What is the right supply chain for your product? *Harvard Business Review* **75**: 105–116.

Huang, S. H., S. K. Sheroran, and H. Keskar. 2005. Computer-aided supply chain configuration based on supply chain operations reference (SCOR) model. *Computers & Industrial Engineering* **48**: 377–394.

Huang, S. H., S. K. Sheoran, and G. Wang. 2004. A review and analysis of supply chain operation reference (SCOR) model. *Supply Chain Management: An International Journal* **9**: 23–29.

Huang, S. H., M. Uppal, and J. Shi. 2002. A product driven approach to manufacturing supply chain selection. *Supply Chain Management: An International Journal* **7**: 189–199.

Mason-Jones, R., B. Naylor, and D. R. Towill. 2000. Lean, agile, or leagile? Matching your supply chain to the marketplace. *International Journal of Production Research* **38**: 4061–4070.

SCC. 2012. Supply chain operations reference (SCOR®) model overview—Version 10.0. Supply Chain Council. http://supply-chain.org/f/SCOR-Overview-Web.pdf (accessed October 12, 2012).

Solomon, M. M. 1987. Algorithms for the vehicle routing and scheduling problems with time window constraints. *Operations Research* **35**: 254–265.

Vonderembse, M. A., M. Uppal, S. H. Huang, and J. P. Dismukes. 2006. Designing supply chains: Towards theory development. *International Journal of Production Economics* **100**: 223–238.

Zhou, H., W. C. Benton Jr., D. A. Schilling, and G. W. Milligan. 2011. Supply chain integration and the SCOR model. *Journal of Business Logistics* **32**: 332–344.

2

Understanding Customer Demand: Forecasting

2.1 Overview

Forecasts of future demand provide the basis for supply chain decision making. For an original equipment manufacturer (OEM) or a supplier, customer demand information is needed in order to make a production plan. For a retailer, customer demand information is needed to make replenishment decisions. To forecast demand, one must first identify the factors that influence future demand and then understand the relationships between these factors and future demand. Factors that influence future demand include

- Product characteristics
- Past demand
- Economic condition
- Competition
- Planned marketing efforts
- Planned price discount

Understanding the impact of these factors is essential to selecting an appropriate forecasting method. Forecasting methods can be classified into four categories: (1) qualitative, (2) causal, (3) time series, and (4) simulation. Qualitative forecasting methods rely on human judgment and hence are subjective. A typical qualitative forecasting method is market survey. First, a questionnaire is developed that contains questions whose answers provide information needed for forecasting. Then, the survey is carried out by distributing the questionnaire to carefully chosen customers through appropriate channels. The responses are then analyzed using statistical tools and the results extrapolated to the general population to come up with a forecast. Details for carrying out a market survey can be found in Kress and Snyder (1994).

Another qualitative forecasting method is expert opinion. Sales and marketing personnel are good examples of "experts" for forecasting the demand of a new product. A formal form of expert opinion is the Delphi technique, named after the Oracle at Delphi who predicted future events in Greek mythology. A facilitator identifies a number of experts that may have different backgrounds to form a committee that corresponds to the Oracle. A questionnaire is developed where the committee members are asked to provide an anonymous forecast of specific events, along with the reasons for making such a forecast. The responses are summarized by the facilitator, and the questionnaire is updated to reflect the results. The updated questionnaire is then returned to the committee members for a second round of response. This process continues until a reasonable agreement among the committee members is reached. The results are then used for decision making. Details for using the Delphi technique in forecasting can be found in Rowe and Wright (1999).

Qualitative forecasting methods are suitable under the following circumstances:

- Little historical data are available. For example, when a new product is introduced.
- Experts have market intelligence that may affect the forecast. For example, a competitor is introducing a new product model.

Causal forecasting methods assume that the demand is highly correlated with certain factors. In this case, regression analysis is used for forecasting. The demand is treated as the dependent variable, whereas the factors are treated as independent variables. For example, demand for umbrella in an amusement park is highly correlated with weather conditions. One may collect umbrella sales data and rainfall data over the past months. A regression model can then be developed to establish the relationship between umbrella demand and rainfall. Given the forecasted rainfall in the future month, one can use the relationship to forecast the demand for umbrella in the same month. Note that in this example there is no time lag between umbrella demand and rainfall. Therefore, we need to use the forecast for rainfall, which may introduce additional uncertainty in the forecast for umbrella demand. If there is time lag between the dependent variable and the independent variable, the regression method will be even more useful.

Time series forecasting methods assume that past demand is a good indicator of future demand. These methods are most appropriate when the demand pattern does not vary significantly from year to year. For example, the demand pattern for consumer staple products such as toilet papers does not vary significantly from year to year when economic condition changes. Therefore, historical demand data can be used to forecast the future demand. These methods are simple to implement and can serve as a good starting point for a demand forecast. Details of these methods will be discussed later.

Simulation forecasting methods mimic consumer behaviors that give rise to demand to arrive at a forecast. Using simulation, one can combine causal and time series methods for forecasting while introducing stochastic elements to represent uncertainty in the real world. Simulation forecasting allows the investigation of various what-if scenarios and hence may produce a better forecast. However, building a simulation model is more time consuming than using other forecasting methods.

Forecasts have the following characteristics:

- Forecasts are always wrong and thus should include an error analysis.
- Long-term forecasts are usually less accurate than short-term forecasts.
- Aggregate forecasts are usually more accurate than disaggregate forecasts.

In general, the farther up the supply chain a company is, the greater is the distortion of information it receives. A classic example is the bullwhip effect (Lee et al. 1997), in which fluctuations in orders increase as they move up the supply chain from retailers to wholesalers to manufacturers to suppliers. As a result, the further up the supply chain a company is, the larger is the forecast error. Such a company should consider collaborating with its supply chain partners to develop forecasts based on sales to the end customer to reduce forecast error.

2.2 Time Series Forecasting

A time series is a time-ordered list of historical data. Time series models include constant, trend, and seasonal. For each model, there are several forecasting methods available, such as average, moving average, exponential smoothing, and regression. To select an appropriate model, one needs to study the historical data and understand the underlying process. We will discuss each model separately.

2.2.1 Constant Process

The weekly demand for AAA batteries at Best Value Electronics (BVE) store in Metropolis for the past 52 weeks is shown in Table 2.1. We are asked to forecast the demand for the next 4 weeks. The first thing we should do is to plot the data as shown in Figure 2.1. From the plot, we can see that the demand fluctuates without any trends or patterns. Therefore, we speculate that the underlying process for demand is constant. In a constant process, the demand in time period t, d_t, can be mathematically represented as

$$d_t = a + \varepsilon_t \tag{2.1}$$

TABLE 2.1

Weekly Demand (in Packs) for AAA Batteries at BVE Store

Week	Demand	Week	Demand	Week	Demand	Week	Demand
1	974	14	897	27	1023	40	959
2	980	15	1022	28	966	41	1014
3	1087	16	934	29	903	42	1039
4	1028	17	1041	30	998	43	1046
5	1031	18	959	31	1043	44	987
6	969	19	977	32	995	45	937
7	1067	20	958	33	971	46	1029
8	925	21	1044	34	974	47	1013
9	981	22	901	35	1007	48	905
10	1045	23	1029	36	980	49	991
11	959	24	969	37	1035	50	993
12	944	25	987	38	1030	51	961
13	943	26	1113	39	943	52	1001

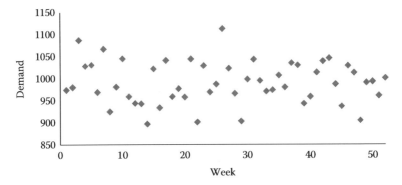

FIGURE 2.1
Plot of weekly demand for AAA batteries at BVE store.

in which a is the underlying constant of the process and ε_t is the random noise. Generally, we assume that ε_t is normally distributed with mean 0 and variance σ_ε^2.

Given historical demand data, our job is to estimate the constant a and the variance of the noise ε_t. Intuitively we can use the average of all past demand to estimate a, that is,

$$\hat{a} = \frac{\sum_{t=1}^{N} d_t}{N} \tag{2.2}$$

in which N is the number of past demand data points. We will show that this estimate actually minimizes the sum of squared errors.

When we use $\hat{a}$ as the estimate of the constant demand, the error in time period t, e_t, is

$$e_t = d_t - \hat{a} \tag{2.3}$$

Assuming we have N data points, to minimize the sum of squared errors, we take its derivative with respect to $\hat{a}$ and set it to 0. We have

$$\frac{d\left(\sum_{t=1}^{N} e_t^2\right)}{d\hat{a}} = \frac{d\left[\sum_{t=1}^{N} (d_t - \hat{a})^2\right]}{d\hat{a}} = -2\sum_{t=1}^{N} (d_t - \hat{a}) = 0 \tag{2.4}$$

$$\sum_{t=1}^{N} d_t = N\hat{a} \tag{2.5}$$

$$\hat{a} = \frac{\sum_{t=1}^{N} d_t}{N} \tag{2.6}$$

It turns out that our intuition of using the average of past demand actually has a sound mathematical foundation. We can then use the variance of e_t (see Equation 2.3) to estimate the variance of the noise ε_t (see Equation 2.1). In this example, we estimate that the weekly demand for AAA batteries for the next 4 weeks is expected to be 991 packs per week. We also estimate that our forecast has a standard deviation of 47.8 (or a variance of 2283.8).

Rather than taking the average of all data points, we can choose to average only the most recent data points. This method is known as the *moving average* method. Let n be the number of data points that we want to consider. The moving average at time period t is calculated as

$$M_t = \frac{\sum_{k=t-n+1}^{t} d_k}{n} \tag{2.7}$$

If we choose a 5-week moving average, that is, $n = 5$, at the end of week 52, the moving average of AAA battery demand at BVE is $M_{52} = (905 + 991 + 993 + 961 + 1001)/5 = 970$. Because we are using a constant model, $M_{52} = 970$ will be used as the estimated weekly demand for AAA batteries for the next 4 weeks, that is, $\hat{d}_{53} = \hat{d}_{54} = \hat{d}_{55} = \hat{d}_{56} = 970$. Note that at the end of week 53, we would know its actual demand, and we can calculate a new moving average. Suppose $d_{53} = 984$, the new moving average in week 53 is $M_{53} = (991 + 993 + 961 + 1001 + 984)/5 = 986$. We can then update the demand forecast for the next 4 weeks as 986, that is, $\hat{d}_{54} = \hat{d}_{55} = \hat{d}_{56} = \hat{d}_{57} = 986$.

There is a more efficient way to calculate a new moving average when a new data point is obtained. The equation is derived as follows:

$$M_{t+1} = \frac{\sum_{k=t+1-n+1}^{t+1} d_k}{n} = \frac{\sum_{k=t-n+1}^{t} d_k + d_{t+1} - d_{t-n+1}}{n}$$

$$= \frac{\sum_{k=t-n+1}^{t} d_k}{n} + \frac{d_{t+1} - d_{t-n+1}}{n}$$

$$= M_t + \frac{d_{t+1} - d_{t-n+1}}{n} \tag{2.8}$$

Therefore, M_{53} can also be calculated as $M_{53} = 970 + ((984 - 905)/5) = 986$.

The effect of moving average is to smooth out noise. The larger the number of periods used in the moving average calculation, n, the smoother the moving average will be. In other words, when using a large n, the moving average forecast is relatively unaffected by noise. Figure 2.2 plots the demand data shown in Table 2.1, along with the 5-and 10-week moving averages. One can see that the 10-week moving average has a smaller variation than the 5-week moving average.

Because the moving average method updates forecast when a new data point is obtained, it is an *adaptive forecasting* method. In other words, the moving average method is capable of responding to changes in the underlying process. The number of periods used in the moving average calculation, n, affects how quickly the forecast will respond to the change in the process. The smaller n is, the faster the forecast will respond to the change. For illustration purpose,

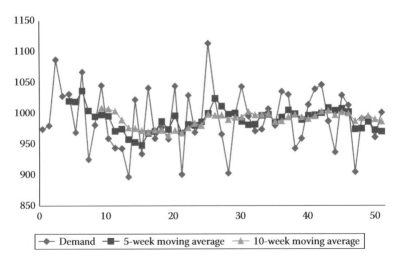

FIGURE 2.2
Comparison of 5- and 10-week moving averages.

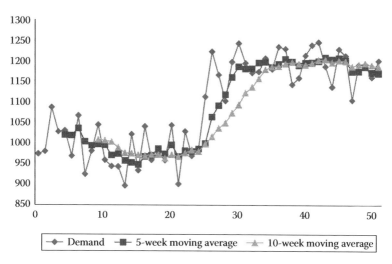

FIGURE 2.3
Comparison of response to process change with different moving average periods.

we add 200 to every data observation after week 26 in Table 2.1, corresponding
to a sudden increase of average demand by 200 packs. Figure 2.3 plots the
new demand data along with the 5- and 10-week moving averages. One can
see that the 5-week moving average responded to the change approximately
5 weeks earlier than the 10-week moving average. Therefore, the selection of
n is a tradeoff between smoothing out noise and quick response to a process
change. If the process is relatively stable, then a larger *n* should be used. On the
other hand, a smaller *n* is better for a process that may be changing. A value of
between 5 and 7 is typically used for *n* in short-term forecasting.

Note that average-based forecasting methods require keeping track of historical data. An alternative method is to use only the most recent forecast
and the new data point to update the forecast. This method is called *exponential smoothing*. Let S_t denote the exponential smoothing estimate at time
period *t*. It can be calculated as

$$S_t = \alpha \times d_t + (1-\alpha) \times S_{t-1} \qquad (2.9)$$

where α is a weighting factor in the range between 0 and 1.

This equation can be expanded as follows:

$$S_t = \alpha \times d_t + (1-\alpha) \times S_{t-1}$$

$$= \alpha d_t + (1-\alpha)\left[\alpha d_{t-1} + (1-\alpha)S_{t-2}\right]$$

$$= \alpha d_t + \alpha(1-\alpha)d_{t-1} + (1-\alpha)^2\left[\alpha d_{t-2} + (1-\alpha)S_{t-3}\right]$$

$$= \alpha d_t + \alpha(1-\alpha)d_{t-1} + \alpha(1-\alpha)^2 d_{t-2} + \cdots + \alpha(1-\alpha)^{t-1}d_1 + (1-\alpha)^t S_0 \quad (2.10)$$

From Equation 2.10, one can see that S_t can be viewed as a "weighted" average of all the data points, with the weight decreasing (because $0 < \alpha < 1$) exponentially with the age of the data point. This is why the method is called exponential smoothing.

Consider the expectation of the exponential smoothing estimate S_t:

$$E[S_t] = E\left[\alpha \sum_{i=0}^{t-1} (1-\alpha)^i d_{t-i} + (1-\alpha)^t S_0\right]$$

$$= E\left[\alpha \sum_{i=0}^{t-1} (1-\alpha)^i d_{t-i}\right] + E\left[(1-\alpha)^t S_0\right] \tag{2.11}$$

Because $0 < \alpha < 1$, we have $\lim_{t\to\infty}(1-\alpha)^t = 0$, so the second term in Equation 2.11 can be dropped. Therefore, we have

$$E[S_t] = E\left[\alpha \sum_{i=0}^{t-1} (1-\alpha)^i d_{t-i}\right] = \alpha \sum_{i=0}^{t-1} (1-\alpha)^i E[d_{t-i}] \tag{2.12}$$

In a constant model, $d_t = a + \varepsilon_t$, where a is a constant and $E[\varepsilon_t] = 0$. Therefore, we have

$$E[d_{t-i}] = E[a + \varepsilon_t] = a \tag{2.13}$$

Because $\lim_{t\to\infty} \alpha \sum_{i=0}^{t-1} (1-\alpha)^i = \alpha/(1-(1-\alpha)) = 1$. We have $E[S_t] = a$. This justifies the use of exponential smoothing as a forecasting method for a constant process.

There are different ways to determine the initial estimate S_{t-1}. The simplest is to average several past data points. Refer to the data shown in Table 2.1, suppose we are at week 10 and simply let $S_9 = d_9 = 981$ and $\alpha = 0.2$. We can calculate $S_{10} = \alpha \times d_{10} + (1 - \alpha) \times S_9 = 0.2 \times 1045 + 0.8 \times 981 = 994$. We then forecast the demand for all subsequent weeks as 994. At week 11, we have $d_{11} = 959$. We then update our forecast as $S_{11} = \alpha \times d_{11} + (1 - \alpha) \times S_{10} = 0.2 \times 959 + 0.8 \times 994 = 987$.

Note that the exponential smoothing method is not very sensitive to the initial estimate. Any discrepancies in the initial estimates will be gradually damped out. Figure 2.4 compares the exponential smoothing results ($\alpha = 0.2$) of different initial estimates starting at week 10 for the data shown in Table 2.1. One can see that the results converge starting at around week 27.

Now we investigate the effect of the weighting factor α. Figure 2.5 plots the results of exponential smoothing using different weighting factor with the same initial estimate. One can see that a smaller α produces a smoother forecast.

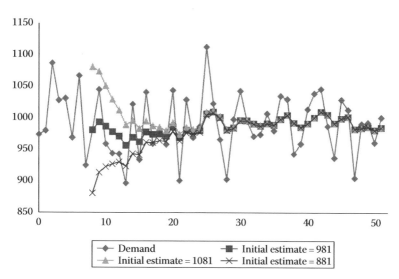

FIGURE 2.4
Comparison of exponential smoothing results using different initial estimates.

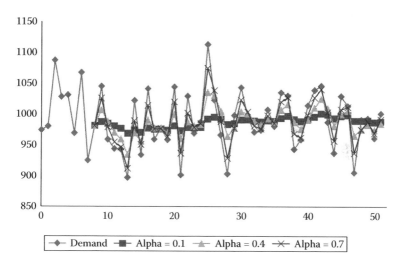

FIGURE 2.5
Comparison of exponential smoothing results using different weighting factors.

The effect is similar to a larger n in the moving average method. Similarly, a smaller α will result in a slower response to the change in the process. Again, we add 200 to every data observation after week 26 and plot the results of exponential smoothing using different weighting factors in Figure 2.6 to illustrate this effect.

The choice of the weighting factor α is a tradeoff between smoothing out noise and quick response. An "optimal" weighting factor can be chosen

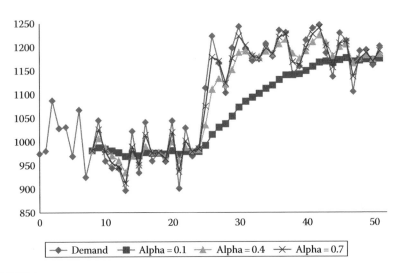

FIGURE 2.6

Comparison of response to process change with different weighting factors.

empirically using the following method. The data are divided into two groups. The first group is used to initialize the forecast procedure. Forecasts are then made for the second group of data using different values of α. The α that produced the most accurate results is then chosen as the "optimal" weighting factor.

2.2.2 Trend Process

Now we look at the monthly demand of laptops at BVE store for the past 24 months, as shown in Table 2.2. Again, we plot the data first, as shown in Figure 2.7. Clearly the underlying process demand is not constant but is steadily increasing. In this case, a model that incorporates a linear trend should be used. In a linear trend model, the demand in time period t can be mathematically represented as

$$d_t = a + bt + \varepsilon_t \tag{2.14}$$

TABLE 2.2

Monthly Demand for Laptops at BVE Store

Month	Demand	Month	Demand	Month	Demand	Month	Demand
1	199	7	217	13	238	19	262
2	210	8	227	14	238	20	259
3	207	9	227	15	244	21	263
4	207	10	227	16	251	22	266
5	218	11	231	17	251	23	266
6	214	12	240	18	252	24	272

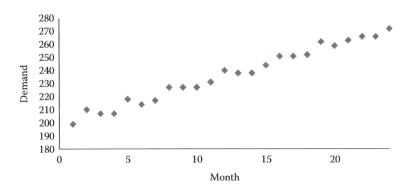

FIGURE 2.7
Plot of monthly demand for laptops at BVE store.

in which a is the level, b is the slope of the trend, and ε_t is the random noise. If b is positive, the trend is increasing; if b is negative, the trend is decreasing.

Now we need to estimate both the level and the slope, given the historical data. This can be done using linear regression, with the time period t as the independent variable and the demand d_t as the dependent variable. Assume we have N data points, the estimate of the slope $\hat{b}$ and the estimate of the level $\hat{a}$ can be derived by minimizing the sum of squared errors. The sum of squared errors is

$$\sum_{t=1}^{N} e_t^2 = \sum_{t=1}^{N} (d_t - \hat{a} - \hat{b}t)^2 \tag{2.15}$$

To minimize this term, we first take its derivative with respect to $\hat{a}$ and set it to 0, as follows:

$$\frac{d\left[\sum_{t=1}^{N} (d_t - \hat{a} - \hat{b}t)^2 \right]}{d\hat{a}} = -2\sum_{t=1}^{N} (d_t - \hat{a} - \hat{b}t) = 0 \tag{2.16}$$

We then take its derivative with respect to $\hat{b}$ and set it to 0, as follows:

$$\frac{d\left[\sum_{t=1}^{N} (d_t - \hat{a} - \hat{b}t)^2 \right]}{d\hat{b}} = -2\sum_{t=1}^{N} (d_t t - \hat{a}t - \hat{b}t^2) = 0 \tag{2.17}$$

Solving the pair of simultaneous Equations 2.16 and 2.17, we have

$$\hat{b} = \frac{N \sum_{t=1}^{N} t d_t - \sum_{t=1}^{N} d_t \sum_{t=1}^{N} t}{N \sum_{t=1}^{N} t^2 - \left(\sum_{t=1}^{N} t \right)^2} \tag{2.18}$$

$$\hat{a} = \frac{\sum_{t=1}^{N} d_t - \hat{b} \sum_{t=1}^{N} t}{N} \tag{2.19}$$

Given the data in Table 2.2, we can use Microsoft Excel to find $\hat{a}$ and $\hat{b}$ using Equations 2.18 and 2.19, as shown in Figure 2.8. We have $\hat{b} = 13.06$ and $\hat{a} = 198.66$.

	A	B	C	D	E	F	G	H
		t	d_t	$t d_t$	t^2			
1								
2		1	199	199	1			
3		2	210	420	4			
4		3	207	621	9			
5		4	207	828	16			
6		5	218	1090	25			
7		6	214	1284	36			
8		7	217	1519	49			
9		8	227	1816	64			
10		9	227	2043	81			
11		10	227	2270	100			
12		11	231	2541	121			
13		12	240	2880	144			
14		13	238	3094	169			
15		14	238	3332	196			
16		15	244	3660	225			
17		16	251	4016	256			
18		17	251	4267	289			
19		18	252	4536	324			
20		19	262	4978	361			
21		20	259	5180	400			
22		21	263	5523	441			
23		22	266	5852	484			
24		23	266	6118	529			
25		24	272	6528	576			
26							$\hat{a}$	$\hat{b}$
27	Total	300	5686	74595	4900		198.66	3.06

Cell	Formula	Note
D2	=B2*C2	Drag down to D25
E2	=B2^2	Drag down to E25
B27	=SUM(B2:B25)	Drag right to E27
H27	=(B25*D27-C27*B27)/(B25*E27-B27^2)	
G27	=(C27-H27*B27)/B25	

FIGURE 2.8
Spreadsheet calculation of slope and level in linear regression.

An easier way is to use Excel to chart the data points first, then add a linear trend line and display the equation on the chart.

When the slope and level are estimated, the linear model can be used to generate future forecast. For example, our forecast for the demand in the next month (time period 25) is $\hat{d}_{25} = 198.66 + 3.06 \times 25 \approx 275$. Our forecast for the demand in month 30 is $\hat{d}_{30} = 198.66 + 3.06 \times 30 \approx 290$.

The regression method requires keeping track of all historical data. We can use the idea of exponential smoothing to update our forecast with a new data point without keeping track of all historical data. Note that this time we need to update both the level and the slope, so the method is called *double exponential smoothing*, which uses the following set of equations:

$$\hat{a}_t = \alpha d_t + (1 - \alpha)(\hat{a}_{t-1} + \hat{b}_{t-1}) \tag{2.20}$$

$$\hat{b}_t = \beta(\hat{a}_t - \hat{a}_{t-1}) + (1 - \beta)\hat{b}_{t-1} \tag{2.21}$$

$$\hat{d}_{t+k} = \hat{a}_t + k\hat{b}_t \tag{2.22}$$

where α and β are weighting factors in the range between 0 and 1. Again, the choice of the weighting factors is a tradeoff between smoothing out noise and quick response.

Give a historical data set from time period 1 to time period t, we need to calculate $\hat{a}_t$ and $\hat{b}_t$ before we can use Equation 2.22 to estimate future demand. The calculation of $\hat{a}_t$ and $\hat{b}_t$ can be done either through direct estimation or by estimating $\hat{a}_0$ and $\hat{b}_0$, followed by the application of Equations 2.20 and 2.21 at each time period until reaching time period t. We look at direct estimation method first. A simple method is to divide the data into two equal groups. If t is an even number, then group 1 contains data points $d_1, d_2, \ldots$ and $d_{\frac{t}{2}}$ and group 2 contains data points $d_{\frac{t}{2}+1}, d_{\frac{t}{2}+2}, \ldots$ and d_t. If t is an odd number, we can drop the first data point so that group 1 contains data points $d_2, d_3, \ldots$ and d_{t+1} and group 2 contains data points $d_{\frac{t+1}{2}+1}, d_{\frac{t+1}{2}+2}, \ldots$ and d_t. We then compute the average of each group and use the difference between these two averages to estimate the slope $\hat{b}_t$. Finally, we estimate the level $\hat{a}_t$ by computing the overall average of the two groups and then add the slope estimate multiplied by the number of time periods between the middle of the data points and time period t. Note that this number equals the number of data points in each group minus 0.5. A graphical illustration of this method is shown in Figure 2.9.

Now we apply this method to the data shown in Table 2.2. First, we divide the data into two groups. Group 1 consists of the demand data from month 1 to month 12; group 2 consists of the demand data from month 13 to month 24. The average of group 1 is 218.67, and the average of group 2 is 255.17. The difference is 36.5. Because the two averages are 12 time periods apart, we divide the difference by 12 and obtain a slope estimate of $\hat{b}_{24} = 3.04$. We then compute

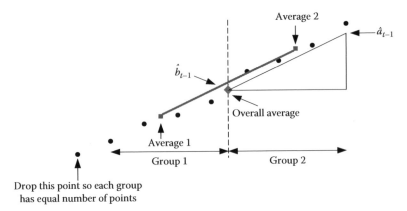

FIGURE 2.9
A simple method to estimate initial level and slope in double exponential smoothing.

the overall average of the two groups, which is 236.92. This average is located at the middle of the data points, which is month 12.5. To bring it to the level at month 24, we need to add the trend adjustment of 3.04 multiplied by 11.5 (which is 24 − 12.5). Therefore, our level estimate is $\hat{a}_{24} = 236.92 + 3.04 \times 11.5 = 271.88$. To estimate the demand in month 25, we use Equation 2.22 and have $\hat{d}_{25} = \hat{d}_{24+1} = \hat{a}_{24} + 1 \times \hat{b}_{24} = 271.88 + 3.04 \approx 275$.

Suppose the actual demand in month 25 is 277 and $\alpha = \beta = 0.2$, we can update our slope and level estimates as follows:

$$\hat{a}_{25} = \alpha d_{25} + (1-\alpha)(\hat{a}_{24} + \hat{b}_{24}) = 0.2 \times 277 + 0.8 \times (271.88 + 3.04) = 275.34$$

$$\hat{b}_{25} = \beta(\hat{a}_{25} - \hat{a}_{24}) + (1-\beta)\hat{b}_{24} = 0.2 \times (275.34 - 271.88) + 0.8 \times 3.04 = 3.12$$

Now, if we want to forecast the demand in month 30, we have $\hat{d}_{30} = \hat{d}_{25+5} = \hat{a}_{25} + 5 \times \hat{b}_{25} = 275.34 + 5 \times 3.12 \approx 291$.

The simple two-group estimation method can also be used to estimate $\hat{a}_0$ and $\hat{b}_0$. The estimation of $\hat{b}_0$ is identical to the estimation of $\hat{b}_t$. To estimate $\hat{a}_0$, we should bring the overall average (at time period 12.5) to the level at time 0 by subtracting the slope multiplied by 12.5. Therefore, $\hat{a}_0 = 236.92 - 3.04 \times 12.5 = 198.92$. Using these values of $\hat{a}_0$ and $\hat{b}_0$ and by applying Equations 2.20 and 2.21 until reaching $t = 24$, we have $\hat{a}_{24} = 271.99$ and $\hat{b}_{24} = 2.97$, which is very close to the values obtained through direct estimation. Similar to exponential smoothing, forecast discrepancies caused by different initial estimates will be gradually damped out.

2.2.3 Seasonal Process

We now look at seasonal demands. The quarterly demand for HDTV at BVE store for the past 5 years is shown in Table 2.3. The data are plotted as shown in Figure 2.10. We can see that the demand from one quarter to the next

TABLE 2.3

Quarterly Demand for HDTV at BVE
Store for the Past 5 Years

Year	Quarter			
	I	II	III	IV
1	434	273	102	228
2	492	353	145	276
3	586	474	146	325
4	717	533	189	354
5	809	632	227	413

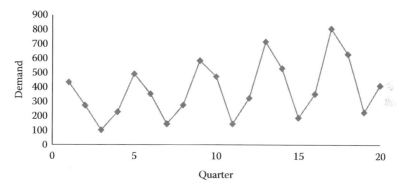

FIGURE 2.10
Plot of quarterly demand for HDTVs at BVE store.

fluctuates quite a bit, whereas the overall trend is increasing. The fluctuation from one quarter to the next is due to the seasonality of the demand. To highlight the seasonality and trend, we plot the demand by quarter for each of the 5 years, as shown in Figure 2.11. For each quarter, the demand increases year over year. This shows that the demand has an increasing trend. In every year, the first quarter demand is the highest; demand declines in the second

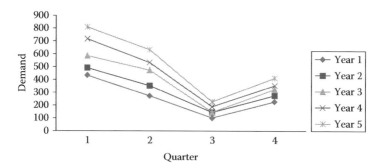

FIGURE 2.11
Plot of HDTV demand by quarter.

quarter and bottoms in the third quarter before recovering in the fourth quarter. This quarterly demand pattern is the seasonality, which is cyclical with each cycle corresponding to a year. There are several ways to model seasonal demand. A popular one is Winters' (1960) multiplicative model, as follows:

$$d_t = (a + bt)c_t + \varepsilon_t \tag{2.23}$$

in which a is the level, b is the slope of the trend, c_t is the seasonal factor for time period t, and ε_t is the random noise.

Here, we use a simple forecasting method that decomposes the seasonal process into two independent components, namely, the seasonal factor and the trend process. The assumptions are (1) the seasonal factor remains unchanged from cycle to cycle and (2) the average of each cycle follows a trend process. Therefore, our method is to first identify the seasonal factor and then remove it from the data. This process is called deseasonalization. We then fit the deseasonalized data using an appropriate trend model. Finally, we use the fitted model to make forecast, taking into account the seasonal factor.

The following steps are used to identify the seasonal factor and deseasonalize the data:

- Determine the average for each cycle
- For each cycle, divide the data value by the average
- Average over each season to compute the seasonal indices
- For each season, divide the original data by the corresponding seasonal index

For the data shown in Table 2.3, each cycle corresponds to a year. Therefore, we calculate the yearly averages; they are 259.25, 316.50, 382.75, 448.25, and 520.25. For each year, we divide the quarterly demand by the yearly average, and then average over each quarter to calculate the seasonal indices. The result is shown in Table 2.4. We can see that the seasonal indices for the four quarters are 1.583, 1.162, 0.418, and 0.837, respectively.

TABLE 2.4

Calculation of Seasonal Indices

Year	Quarter			
	I	II	III	IV
1	1.674	1.053	0.393	0.879
2	1.555	1.115	0.458	0.872
3	1.531	1.238	0.381	0.849
4	1.600	1.189	0.422	0.790
5	1.555	1.215	0.436	0.794
Seasonal index	1.583	1.162	0.418	0.837

TABLE 2.5

Deseasonalized Data

Year	Quarter			
	I	II	III	IV
1	274.19	234.91	243.90	272.45
2	310.84	303.75	346.72	329.81
3	370.22	407.87	349.12	388.37
4	452.99	458.64	451.94	423.02
5	511.11	543.83	542.80	493.52

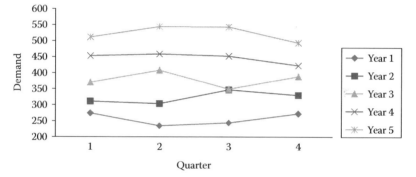

FIGURE 2.12
Plot of deseasonalized HDTV demand by quarter.

For each quarter, we divide the original data by the corresponding seasonal index. The result is deseasonalized data, as shown in Table 2.5.

Now we plot the deseasonalized demand by quarter, as shown in Figure 2.12. We can see that the seasonal pattern no longer exists, whereas the yearly increasing trend remains. We can then fit this deseasonalized data using the following steps:

- Calculate the average for each cycle using the deseasonalized data
- Fit the cycle average data using an appropriate trend model

We calculate the yearly average of the deseasonalized data; they are 256.37, 322.78, 378.89, 446.65, and 522.82. We plot these averages (shown in Figure 2.13) and found that the trend appears to be linear. Therefore, we use linear regression to fit the data and obtain the model $d_t = 188.47 + 65.68t$. Note that the period t represents year. The following steps are then used to make forecast:

- Use the model to forecast the cycle average
- Multiply the cycle average by the seasonal indices to make seasonal forecast

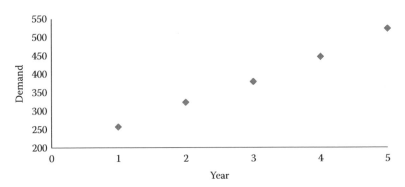

FIGURE 2.13
Plot of yearly average of deseasonalized HDTV demand.

To forecast the demand for year 6, we first forecast the average demand for year 6 as $\hat{d}_6 = 188.47 + 65.68 \times 6 = 582.55$. Therefore, in year 6, the demand for the first quarter is $582.55 \times 1.583 \approx 922$; the demand for the second quarter is $582.55 \times 1.162 \approx 677$; the demand for the third quarter is $582.55 \times 0.418 \approx 244$; and the demand for the fourth quarter is $582.55 \times 0.837 \approx 488$.

2.3 Error Analysis

As previously mentioned, forecast is always wrong, which is due to the random component of the underlying process. Therefore, we need to conduct an error analysis for our forecast. Forecast error (e_t) is the difference between the actual demand (d_t) and the forecast $(\hat{d}_t)$. Mathematically, we have

$$e_t = d_t - \hat{d}_t \tag{2.24}$$

Looking at the error for an isolated period does not provide useful information. Rather, we should study the errors over the history of the entire forecast period in order to evaluate the performance of a forecasting model. There are a number of performance measures. Assume we have used a forecasting model for n periods of time. The *bias* of the model is the sum of the forecast errors, defined as

$$bias = \sum_{t=1}^{n} e_t \tag{2.25}$$

If the model behaves properly, we would expect the bias to be 0 because the random noise of the underlying process, ε_t, follows a normal distribution with zero mean. If the bias of a model deviates significantly from 0, then the model is either underestimating (when the bias is positive) or overestimating (when the bias is negative) the demand.

When the bias of a model hovers around 0, it does not necessarily mean that the model is appropriate. For example, if a model always underestimates the demand in odd period and overestimates the demand by the same amount in even period, its bias will be 0. To detect such a problem, the *mean absolute deviation* (MAD) is often used. It is defined as

$$MAD = \frac{1}{n} \sum_{t=1}^{n} |e_t| \qquad (2.26)$$

MAD measures the dispersion of the errors. A larger MAD value indicates problems with the model.

Brown (1963) showed that for a normal distribution the MAD and the error standard deviation (σ_ε) are related by

$$MAD = \sqrt{\frac{2}{\pi}} \sigma_\varepsilon \qquad (2.27)$$

Because $\sqrt{2/\pi} \approx 0.8$, it is common to estimate the error standard deviation using the following:

$$\sigma_\varepsilon \approx \frac{MAD}{0.8} \qquad (2.28)$$

Note that even for non-normal distributions, MAD/0.8 is still a good approximation of the error standard deviation.

Another measure of error dispersion is the *mean squared error* (MSE), defined as

$$MSE = \frac{1}{n} \sum_{t=1}^{n} e_t^2 \qquad (2.29)$$

By squaring the error, the penalty is increased for larger errors. MSE is the second moment of the error that incorporates both the bias and variance of the forecast model. A regression-based forecasting model aims to minimize its MSE.

Note that both MAD and MSE depend on the magnitude of the demand to be forecasted. If the demand is large, the errors tend to be large and thus

MAD and MSE are large. It may be more meaningful to look at the errors relative to the magnitude of the demand. This is done by using the *mean absolute percentage error* (MAPE), defined as

$$\text{MAPE} = \frac{1}{n}\left(\sum_{t=1}^{n}\frac{|e_t|}{d_t}\times 100\right) \tag{2.30}$$

	A	B	C	D	E	F	G
1	Quarter	d_t	$\hat{d}_t$	e_t	$|e_t|$	e_t^2	$|e_t|/d_t$
2	1	434					
3	2	273					
4	3	102					
5	4	228					
6	5	492	259	233	233	54289	0.47358
7	6	353	274	79	79	6241	0.2238
8	7	145	294	-149	149	22201	1.02759
9	8	276	305	-29	29	841	0.10507
10	9	586	317	269	269	72361	0.45904
11	10	474	340	134	134	17956	0.2827
12	11	146	370	-224	224	50176	1.53425
13	12	325	371	-46	46	2116	0.14154
14	13	717	383	334	334	111556	0.46583
15	14	533	416	117	117	13689	0.21951
16	15	189	430	-241	241	58081	1.27513
17	16	354	441	-87	87	7569	0.24576
18	17	809	448	361	361	130321	0.44623
19	18	632	471	161	161	25921	0.25475
20	19	227	496	-269	269	72361	1.18502
21	20	413	506	-93	93	8649	0.22518
22				Bias	MAD	MSE	MAPE
23				550	176.6	40895.5	53.5311

Cell	Formula	Note
C6	=ROUND(AVERAGE(B2:B5),0)	Drag down to C21
D6	=B6-C6	Drag down to D21
E6	=ABS(D6)	Drag down to E21
F6	=D6*D6	Drag down to F21
G6	=E6/B6	Drag down to G21
D23	=SUM(D6:D21)	
E23	=AVERAGE(E6:E21)	
F23	=AVERAGE(F6:F21)	
G23	=AVERAGE(G6:G21)*100	

FIGURE 2.14
Forecasting result and error analysis for the 4-quarter moving average model.

The *tracking signal* (TS) is commonly used to monitor the randomness of the forecast error. It is the ratio of the bias and the MAD, given as

$$TS_t = \frac{bias_t}{MAD_t} \qquad (2.31)$$

The tracking signal is compared to predefined control limits to determine if the actual demand reflects the assumptions of the forecast model. Commonly used control limits are ±4 or ±6. Because $\sigma_\varepsilon \approx MAD/0.8$, the ±4 control limits are an approximation of the three-signal control limits in statistical process control ($4MAD \approx 3\sigma_\varepsilon$).

Now, let us compare three different models for forecasting the HDTV demand shown in Table 2.3. The first model is a four-quarter moving average model; the second model is a linear trend model ($\hat{b} = 11.38$ and $\hat{a} = 265.8$); and the third model is the seasonal model we previously developed. We use Microsoft Excel for the calculations. The results for the three models are shown in Figures 2.14 through 2.16, respectively. Note that when using the four-quarter moving average model we have no forecast until the fifth quarter, when error analysis can start. For comparison purpose, the error analysis for the trend and seasonal models also starts from the fifth quarter. From the error analysis results, it is obvious that the seasonal model has the best performance.

2.4 Case Studies

Case Study 2.4.1

BVE store starts selling smart phones a year ago. The historical sales data are shown in Table 2.6. You asked two student interns, John and Mary, to develop a forecast model. John believed that the sales had an increasing trend and developed a trend model $\hat{d}_t = 96.15 + 0.45t$. Mary thought that the sales are more or less constant and thus presented a constant model $\hat{d}_t = 99$. Both of them conducted error analysis, and the results are shown in Table 2.7. Whose model should you use?

From Table 2.7, we can see that both models perform well, but John's trend model appears to be better (having lower MAD, MSE, and MAPE). However, the error analysis was conducted based on historical data, which were used for developing the model. Therefore, the conclusion we can draw is that John's model fits the historical data slightly better. This is because the trend model is more complicated than the constant model, and hence fits the data better. However, we cannot say definitely that the trend model will provide a more accurate forecast. Therefore, we decide to keep both models and validate them using data collected in the next 6 months.

The actual sales of smart phones in the subsequent 6 months are 96, 98, 102, 100, 109, and 100, respectively. We conduct error analysis on both John and Mary's models using this validation data set. The results are

	A	B	C	D	E	F	G
1	Quarter	d_t	$\hat{d}_t$	e_t	$|e_t|$	e_t^2	$|e_t|/d_t$
2	1	434					
3	2	273					
4	3	102					
5	4	228					
6	5	492	323	169	169	28561	0.3435
7	6	353	334	19	19	361	0.05382
8	7	145	345	-200	200	40000	1.37931
9	8	276	357	-81	81	6561	0.29348
10	9	586	368	218	218	47524	0.37201
11	10	474	380	94	94	8836	0.19831
12	11	146	391	-245	245	60025	1.67808
13	12	325	402	-77	77	5929	0.23692
14	13	717	414	303	303	91809	0.42259
15	14	533	425	108	108	11664	0.20263
16	15	189	437	-248	248	61504	1.31217
17	16	354	448	-94	94	8836	0.26554
18	17	809	459	350	350	1E+05	0.43263
19	18	632	471	161	161	25921	0.25475
20	19	227	482	-255	255	65025	1.12335
21	20	413	493	-80	80	6400	0.1937
22				Bias	MAD	MSE	MAPE
23				142	168.9	36966	54.7675

Cell	Formula	Note
C6	=ROUND(11.38*A6+265.8,0)	Drag down to C21
D6	=B6-C6	Drag down to D21
E6	=ABS(D6)	Drag down to E21
F6	=D6*D6	Drag down to F21
G6	=E6/B6	Drag down to G21
D23	=SUM(D6:D21)	
E23	=AVERAGE(E6:E21)	
F23	=AVERAGE(F6:F21)	
G23	=AVERAGE(G6:G21)*100	

FIGURE 2.15
Forecasting result and error analysis for the trend model.

shown in Table 2.8. Now it appears that Mary's model performs better on the validation data.

We then collected another 6 months of actual sales data. They are 103, 99, 92, 95, 92, and 108. We calculate the tracking signals for both John and Mary's model using these 12 months of data. The results are shown in Table 2.9 and plotted in Figure 2.17. We can see that the tracking signals from Mary's constant model fluctuate around 0, with a highest value of 5.22. On the other hand, the tracking signals from John's trend

	A	B	C	D	E	F	G	H
1	Quarter	Seasonal Index	d_t	$\hat{d}_t$	e_t	$|e_t|$	e_t^2	$|e_t|/d_t$
2	1	1.583	434					
3	2	1.162	273					
4	3	0.418	102					
5	4	0.837	228					
6	5	1.583	492	506	-14	14	196	0.02846
7	6	1.162	353	372	-19	19	361	0.05382
8	7	0.418	145	134	11	11	121	0.07586
9	8	0.837	276	268	8	8	64	0.02899
10	9	1.583	586	610	-24	24	576	0.04096
11	10	1.162	474	448	26	26	676	0.05485
12	11	0.418	146	161	-15	15	225	0.10274
13	12	0.837	325	323	2	2	4	0.00615
14	13	1.583	717	714	3	3	9	0.00418
15	14	1.162	533	524	9	9	81	0.01689
16	15	0.418	189	189	0	0	0	0
17	16	0.837	354	378	-24	24	576	0.0678
18	17	1.583	809	818	-9	9	81	0.01112
19	18	1.162	632	601	31	31	961	0.04905
20	19	0.418	227	216	11	11	121	0.04846
21	20	0.837	413	433	-20	20	400	0.04843
22					Bias	MAD	MSE	MAPE
23					-24	14.13	278.3	3.98597

Cell	Formula	Note
D6	=ROUND((188.47+65.68*ROUND(A6/4+0.25,0))*B6,0)	Drag down to C21
E6	=C6-D6	Drag down to D21
F6	=ABS(E6)	Drag down to E21
G6	=E6*E6	Drag down to F21
H6	=F6/C6	Drag downto G21
E23	=SUM(E6:E21)	
F23	=AVERAGE(F6:F21)	
G23	=AVERAGE(G6:G21)	
H23	=AVERAGE(H6:H21)*100	

FIGURE 2.16
Forecasting result and error analysis for the seasonal model.

model are consistently below 0, have a decreasing trend, and go below −6 after the 20th month. It is obvious that John's model overestimates the actual demand. The discrepancy between John's estimate and the actual demand gets larger and larger as time goes by. Apparently the demand of smart phones follows a constant process. The small increasing trend (a slope of 0.45) that John derived from the first 12 months of data was due to the random fluctuation of the constant process.

In general, a more complex model will provide a better fit of the data, resulting in better performance in terms of error analysis. However, the complex model may be overfitting the data. Therefore, model validation (i.e., conducting error analysis using data points that are not used in model building) is very important.

TABLE 2.6

Historical Sales Data of Smart Phones at BVE Store

Month	1	2	3	4	5	6	7	8	9	10	11	12
Sales	97	95	96	98	105	95	100	101	101	99	100	102

TABLE 2.7

Results of Error Analysis for the Two Models Using Historical Data

	Bias	MAD	MSE	MAPE
Trend model	−1	1.75	6.75	1.76
Constant model	1	2.42	8.42	2.44

TABLE 2.8

Results of Error Analysis for the Two Models Using the Validation Data

	Bias	MAD	MSE	MAPE
Trend model	−13	3.83	17.17	3.82
Constant model	11	3.17	20.17	3.04

TABLE 2.9

Tracking Signal Calculation for the Two Models

Month	Sales	Trend Model				Constant Model			
		Forecast	Bias	MAD	TS	Forecast	Bias	MAD	TS
13	96	102	−6	6.00	−1.00	99	−3	3.00	−1.00
14	98	102	−10	5.00	−2.00	99	−4	2.00	−2.00
15	102	103	−11	3.67	−3.00	99	−1	2.33	−0.43
16	100	103	−14	3.50	−4.00	99	0	2.00	0.00
17	109	104	−9	3.80	−2.37	99	10	3.60	2.78
18	100	104	−13	3.83	−3.39	99	11	3.17	3.47
19	103	105	−15	3.57	−4.20	99	15	3.29	4.57
20	99	105	−21	3.88	−5.42	99	15	2.88	5.22
21	92	106	−35	5.00	−7.00	99	8	3.33	2.40
22	95	106	−46	5.60	−8.21	99	4	3.40	1.18
23	92	107	−61	6.45	−9.45	99	−3	3.73	−0.80
24	108	107	−60	6.00	−10.00	99	6	4.17	1.44

Case Study 2.4.2

You now asked John to look into the monthly GPS sales data from the past 2 years (column B in Figure 2.18). John thought that the demand follows a constant process but the fluctuation of sales from month to month is quite large. He decided to use an exponential smoothing model to

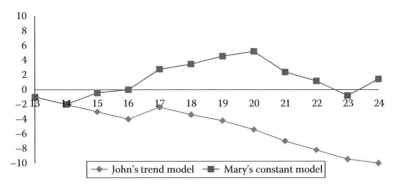

FIGURE 2.17
Plot of tracking signals for the two models.

	A	B	C	D	E	F	G	H	I	J	K	L
1					α					MAPE		
2	Month	Demand	0.1	0.2	0.3	0.4	0.5	α=0.1	α=0.2	α=0.3	α=0.4	α=0.5
3	1	1059										
4	2	1136										
5	3	900										
6	4	942										
7	5	956										
8	6	1094										
9	7	1032	1015	1015	1015	1015	1015	0.016473	0.016473	0.016473	0.016473	0.016473
10	8	911	1017	1018	1020	1022	1024	0.116356	0.117453	0.119649	0.121844	0.12404
11	9	970	1006	997	987	978	968	0.037113	0.027835	0.017526	0.008247	0.002062
12	10	934	1002	992	982	975	969	0.072805	0.062099	0.051392	0.043897	0.037473
13	11	1040	995	980	968	959	952	0.043269	0.057692	0.069231	0.077885	0.084615
14	12	1026	1000	992	990	991	996	0.025341	0.033138	0.035088	0.034113	0.02924
15	13	969	1003	999	1001	1005	1011	0.035088	0.03096	0.033024	0.037152	0.043344
16	14	980	1000	993	991	991	990	0.020408	0.013265	0.011224	0.011224	0.010204
17	15	968	998	990	988	987	985	0.030992	0.022727	0.020661	0.019628	0.017562
18	16	959	995	986	982	979	977	0.037539	0.028154	0.023983	0.020855	0.01877
19	17	1041	991	981	975	971	968	0.048031	0.057637	0.063401	0.067243	0.070125
20	18	1023	996	993	995	999	1005	0.026393	0.029326	0.02737	0.02346	0.017595
21	19	999	999	999	1003	1009	1014	0	0	0.004004	0.01001	0.015015
22	20	980	999	999	1002	1005	1007	0.019388	0.019388	0.022449	0.02551	0.027551
23	21	1063	997	995	995	995	994	0.062088	0.06397	0.06397	0.06397	0.064911
24	22	1016	1004	1009	1015	1022	1029	0.011811	0.00689	0.000984	0.005906	0.012795
25	23	896	1005	1010	1015	1020	1023	0.121652	0.127232	0.132813	0.138393	0.141741
26	24	1049	994	987	979	970	960	0.052431	0.059104	0.06673	0.07531	0.084843
27												
28								3.881836	3.822104	3.92178	4.155509	4.37046

Cell	Formula	Note
C9	=ROUND(AVERAGE($B$3:$B$8),0)	Drag right to G9
C10	=ROUND($B9*C$2+C9*(1-C$2),0)	Drag right to G10; then drag down to G26 with C10 to G10 selected
H9	=ABS($B9-C9)/$B9	Drag right to L9; then drag down to L26 with H9 to L9 selected
H28	=AVERAGE(H15:H26)*100	Drag right to L28

FIGURE 2.18
Finding an appropriate weighting factor for an exponential smoothing model.

smooth out the noise. He used the average of the first 6 months of sales as the initial estimate and tried five different weighting factors ($\alpha = 0.1$, 0.2, 0.3, 0.4, 0.5). He then calculated the MAPE from the 7th month to the 24th month. The results are shown in Figure 2.18. It was found that the MAPE is the lowest (at 3.82) when $\alpha = 0.2$. Therefore, John recommended using an exponential smoothing model with $\alpha = 0.2$ for forecasting. In the next 12 months, the exponential smoothing model does not seem to work well. The actual demand is shown in column B in Figure 2.19. You calculated the tracking signals as shown in column H in Figure 2.19. What is going on?

From Figure 2.19, we can see that the tracking signals keep increasing and exceed 6 after the 31st month. Apparently the assumption of a constant process no longer holds. We plot the actual demand and the forecast as shown in Figure 2.20. We can see that the actual demand appears to be increasing. Although the exponential smoothing forecast also increases, it lags the increase of actual demand. Because the demand now follows a trend process (perhaps due to the increased popularity of GPS more and more people are buying them), we should change our forecast model to a trend model (e.g., a double exponential smoothing model).

	A	B	C	D	E	F	G	H
1	Month	Demand	Forecast	e	|e|	Bias	MAD	TS
2	24	1049	987					
3	25	1040	999	41	41	41	41	1
4	26	1013	1007	6	6	47	23.5	2
5	27	992	1008	-16	16	31	21	1.47619
6	28	1132	1005	127	127	158	47.5	3.326316
7	29	1196	1030	166	166	324	71.2	4.550562
8	30	1039	1063	-24	24	300	63.33333	4.736842
9	31	1181	1058	123	123	423	71.85714	5.88668
10	32	1131	1083	48	48	471	68.875	6.838475
11	33	1186	1093	93	93	564	71.55556	7.881988
12	34	1247	1112	135	135	699	77.9	8.973042
13	35	1250	1139	111	111	810	80.90909	10.01124
14	36	1231	1161	70	70	880	80	11
15								

Cell	Formula	Note
C3	=ROUND(B2*0.2+0.8*C2,0)	Drag down to C14
D3	=B3-C3	Drag down to D14
E3	=ABS(D3)	Drag down to E14
F3	=SUM(D$3:D3)	Drag down to F14
G3	=AVERAGE(E$3:E3)	Drag down to G14
H3	=F3/G3	Drag down to H14

FIGURE 2.19
Analysis of the exponential smoothing model.

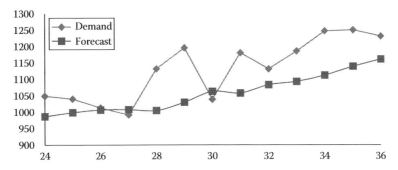

FIGURE 2.20
Plot of actual GPS demand versus forecast.

Case Study 2.4.3

C & K, Inc. is a luxury watch maker. Its watches are exclusively distributed by Western Trading Company. The monthly sales to Western Trading Company over the past 2 years follow a constant process but have high fluctuation. The monthly sales are as low as 330 and as high as 1813. This creates a problem for production scheduling at C & K. Mark Kerr, the CEO at C & K, does not believe that the end customer demand has such a high level of fluctuation. He discussed his concerns with the purchasing manager at Western Trading Company. It turns out that unmet demand in a particular month at Western Trading Company is backlogged and satisfied the next month. Therefore, Western Trading Company uses a simple method for ordering C & K watches based on observed customer demand and inventory/backlog on hand. It always uses the current month customer demand as the forecast for next month's customer demand. If there is inventory at the end of the month, its order is the forecast minus the inventory. If there is backlog at the end of the month, its order is the forecast plus the backlog. Western Trading Company provided the inventory and backlog information to C & K. These data, along with data of sales to Western Trading Company over the past 2 years, are shown in Table 2.10. Is Mark's belief justified?

First, we use C & K sales data to develop a forecast and analyze the error. Because the sales follow a constant process, we use Equation 2.2, that is, the average over the past 2 years, to forecast the monthly sales and have $\hat{a} = 951$. Using Equations 2.24 and 2.26, we can calculate the MAD as 286.58. Using Equation 2.28, we estimate the error standard deviation as $\sigma_\varepsilon = 358$.

Now we use the inventory and backlog data at Western Trading Company along with the C & K sales data to determine the actual monthly customer demand of C & K watches. Let S_t, I_t, and B_t denote the C & K sales, inventory at Western Trading Company, and backlog at Western Trading Company in month t, respectively. Based on Western Trading Company's ordering policy, we can calculate the actual monthly customer demand as

TABLE 2.10

C & K Watch Sales and Inventory/Backlog Data
from Western Trading Company

Month	Sales	Inventory	Backlog	Month	Sales	Inventory	Backlog
1	1000	46	0	13	800	0	133
2	908	99	0	14	1138	44	0
3	756	15	0	15	917	225	0
4	825	0	326	16	511	0	462
5	1492	101	0	17	1660	434	0
6	964	183	0	18	330	0	184
7	699	48	0	19	1132	132	0
8	786	0	147	20	684	41	0
9	1128	0	416	21	734	0	312
10	1813	489	0	22	1399	0	55
11	419	0	36	23	1197	292	0
12	980	72	0	24	558	0	26

TABLE 2.11

Actual Monthly Demand of C & K Watches

Month	Demand	Month	Demand	Month	Demand	Month	Demand
1	954	7	834	13	1005	19	816
2	855	8	981	14	961	20	775
3	840	9	1397	15	736	21	1087
4	1166	10	908	16	1198	22	1142
5	1065	11	944	17	764	23	850
6	882	12	872	18	948	24	876

$d_t = S_t + I_{t-1} - B_{t-1} - I_t + B_t$, shown in Table 2.11. Using Equation 2.2 on these data, we have $\hat{a} = 952$. Using Equations 2.24, 2.26, and 2.28, we have $\sigma_\varepsilon = 149$.

We can see that the forecasts using sales data and actual demand data are nearly identical. However, the error standard deviation obtained using the actual demand data is much smaller than that obtained using the sales data. Therefore, Mark's belief is justified. Although Western Trading Company has a rational ordering policy, the fluctuation of sales observed by C & K is much higher than the fluctuation of actual demand from end customers, as shown in Figure 2.21. This is a manifestation of the bullwhip effect previously mentioned. Mark Kerr realized that sales do not equal actual customer demand. By working with Western Trading Company, a company closer to the end customer in the supply chain, C & K can obtain more accurate demand data. Lower error standard deviation of the forecast means that less flexibility is required for C & K production scheduling. C & K can also work with Western Trading Company to determine a more leveled monthly ordering plan to simplify its production scheduling while maintaining customer satisfaction.

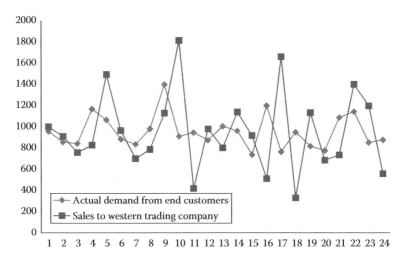

FIGURE 2.21
Comparison of sales observed at C & K and actual demand from end customers.

Case Study 2.4.4

Tony's Lamb Holuse is a distributor of gourmet lamb shanks. It imports lamb shanks at the beginning of every quarter from Australia, New Zealand, and China. It currently rents a warehouse with a capacity to store 200,000 lb of lamb shanks. The demand for lamb shanks has increased steadily over the past 3 years, as shown in Table 2.12.

TABLE 2.12

Demand of Lamb Shank (in lb) at Tony's Lamb House

Quarter	Australia	New Zealand	China	Total
1	26,734	9,678	40,066	76,478
2	24,238	14,645	43,866	82,749
3	25,492	18,696	47,491	91,679
4	28,978	21,277	49,236	99,491
5	27,868	28,498	61,772	118,138
6	28,620	31,646	60,853	121,119
7	29,455	46,087	53,756	129,298
8	36,785	39,703	55,506	131,994
9	40,562	51,824	63,195	155,581
10	39,778	55,854	59,741	155,373
11	39,007	56,438	59,897	155,342
12	47,682	62,439	62,518	172,639

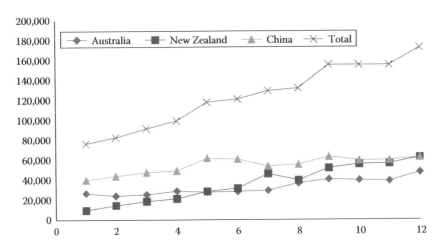

FIGURE 2.22
Plot of lamb shank demand at Tony's Lamb House.

A market study shows that the demand is likely to keep increasing for another 3 years and then stabilize. The current warehouse capacity will not be sufficient for storage in the future. Therefore, Tony's Lamb House decides to build its own warehouse. The capacity of the warehouse should be 110% of the forecasted demand. How big a warehouse should Tony's Lamb House build? When should the warehouse be completed by?

First, we plot the demand data as shown in Figure 2.22. We can see that the demands of lamb shanks from different countries have different rate of growth. Because all three sources of lamb shanks will be stored in a single warehouse, it is not necessary to develop three different forecasts. Rather, we can simply forecast the total demand. We use a linear trend model to fit the total demand data and have $\hat{d}_t = 67{,}630 + 8{,}696t$. Because the demand is expected to stabilize after 3 years, we should use the forecasted demand in the 24th quarter to determine the warehouse capacity. We have $\hat{d}_{24} = 67{,}630 + 8{,}696 \times 24 = 276{,}334$. Therefore, Tony's Lamb House should build a warehouse with a capacity of storing $276{,}334 \times 110\% = 303{,}967$ lb of lamb shank. Note that if three separate linear trend models are developed and the forecasts combined, the result will be the same.

Using the linear trend model, we can forecast the demand from the 13th quarter to the 16th quarter as 180,678, 189,374, 198,070, and 206,766 lb, respectively. Using the data from the first 12 quarters and the linear trend model, we calculate the MAD as 3400 and the error standard deviation σ_ε as 4250. Assuming the total demand follows a normal distribution with a constant standard deviation equals the error standard deviation, there is a 95% probability that the total demand in the 15th quarter will exceed $200{,}000$ lb $(198{,}070 + F_S^{-1}(0.95) \times 4{,}250 = 198{,}070 + 1.645 \times 4{,}250 = 205{,}061)$. Therefore, Tony's Lamb House should plan to complete building the warehouse within three quarters.

Problem: Oriental Trading Company

Oriental Trading Company (OTC) imports rice from Thailand and repackages it for sale. Rice is imported every quarter and stored in a public warehouse before selling to various supermarkets. The public warehouse charges customers for both material handling and storage. Material handling cost is $10 for every thousand pounds of rice. Storage cost is $50 per thousand pounds of rice in storage per quarter. For example, if OTC imported 500,000 lb of rice at the beginning of the quarter and sold all of the rice at the end of the quarter, it would have incurred $10,000 of material handling cost ($5,000 for unloading at the warehouse for storage and $5,000 for loading from the warehouse to sell to its customers) and $12,500 ((500 + 0)/2 × $50) of storage cost.

OTC's major customer is Mega Mart, which accounts for about 75% of its business. Sales to Mega Mart have increased steadily over the past 5 years, although with seasonal fluctuations. The other 25% of the sales are distributed across a number of smaller supermarkets with no apparent seasonality. The sales data from the past 5 years are shown in Table 2.13.

TABLE 2.13

Rice Sale (in 1000 lb) at Oriental Trading Company over the Past 5 Years

Year	Quarter	Sales to Mega Mart	Sales to Other Super Markets	Total Sales
2008	I	510	206	716
	II	561	274	835
	III	1008	255	1263
	IV	488	223	711
2009	I	545	236	781
	II	588	237	825
	III	1145	228	1373
	IV	651	266	917
2010	I	520	249	769
	II	586	266	852
	III	1203	257	1460
	IV	563	261	824
2011	I	592	238	830
	II	604	289	893
	III	1268	223	1491
	IV	648	262	910
2012	I	627	241	838
	II	648	261	909
	III	1358	292	1650
	IV	670	251	921

OTC is now considering the option of leasing a private warehouse for its business needs. On average, 1 sq. ft. of space is required to store 1000 lb of rice, taking into account aisle space. Leasing a private warehouse requires a 3-year contract. The cost is $100 per square feet per year, with a minimum size of 900 ft². Therefore, leasing a private warehouse requires a minimum of $270,000 of capital investment. Operating a private warehouse also requires material handling, but the cost is only $6 for every thousand pounds of rice.

OTC is aware of the bullwhip effect in supply chain. It has worked with Mega Mart to understand its procurement decisions. Every year Mega Mart makes a forecast and adjusts its order quantity based on actual demand. It backlogs demands during stockout by offering customers free delivery and adds the amount of the backlog to next quarter's forecast. If there is excess inventory at the end of a quarter, it subtracts the excess amount from next quarter's forecast. The actual customer demand at Mega Mart is shown in Table 2.14.

TABLE 2.14

Customer Demand of Rice (in 1000 lb)
at Mega Mart over the Past 5 Years

Year	Quarter	Demand
2008	I	551
	II	468
	III	1008
	IV	535
2009	I	578
	II	565
	III	1211
	IV	510
2010	I	576
	II	583
	III	1163
	IV	582
2011	I	594
	II	608
	III	1288
	IV	617
2012	I	638
	II	658
	III	1350
	IV	666

OTC hired you as a consultant. You need to prepare a report that includes the following:

- Demand forecast for the next 3 years.
- Warehouse strategy. Should OTC continue using the public warehouse? Or should it lease a private warehouse? How big a private warehouse is needed? Should OTC use a combination of public and private warehouses?

Exercises

2.1 Given the following demand data, use exponential smoothing to forecast the demand in time period 13:

Time	1	2	3	4	5	6	7	8	9	10	11	12
Demand	105	82	111	109	94	99	117	116	103	96	89	103

   (a) Use the average over the 12 time periods as the initial estimate (S_0) and try different α values (from 0.1 to 0.9 with an increment of 0.1). Which α value gives the best fit to the demand data?

   (b) Use different values for the initial estimate (e.g., average over the first 3, 6, and 9 time periods) and reevaluate the α value that gives the best fit to the demand data. Is this α value influenced by the initial estimate?

2.2 Consider the following data for airplane sales from 2006 to 2012. Use a 3-year moving average model and a linear trend model to forecast the sales in 2013. Which model is likely to give you a better result?

2006	2007	2008	2009	2010	2011	2012
15	20	35	40	55	70	80

2.3 Consider the following noise-free data generated from the linear trend model $d_t = 2 + 3t$. Show that using the method illustrated in Figure 2.9 to estimate the initial slope and level, a double exponential smoothing method will provide noise-free forecast no matter what α and β values are used.

Time	1	2	3	4	5	6	7	8	9	10	11	12
Demand	5	8	11	14	17	20	23	26	29	32	35	38

2.4 Quarterly sales of air-conditioning units at Best Value Appliances for the past 3 years are as follows. Forecast the quarterly sales in 2013 using Winters' multiplicative model. Compare the result by simply averaging quarterly sales over the 3 years.

Year	I	II	III	IV
	Quarter			
2010	100	125	240	100
2011	95	125	240	105
2012	105	120	240	100

2.5 Forecast current year quarterly motorcycle sales given data from the past 5 years shown in the following table. Suppose the actual quarterly sale during the current year is 140, 40, 35, and 80, what is the behavior of the tracking signal based on your model? What does this say about your model?

Year	I	II	III	IV
	Quarter			
1	40	12	8	20
2	60	15	15	30
3	80	20	20	40
4	100	20	30	50
5	120	30	30	60

References

Brown, R. G. 1963. *Smoothing, Forecasting and Prediction of Discrete Time Series.* Englewood Cliffs, NJ: Prentice Hall.

Kress, G. J. and J. Snyder. 1994. *Forecasting and Market Analysis Techniques: A Practical Approach.* Westport, CT: Quorum Books.

Lee, H. L., V. Padmanabhan, and S. Whang. 1997. The bullwhip effect in supply chains. *Sloan Management Review* **38**: 93–102.

Rowe, G. and G. Wright. 1999. The Delphi technique as a forecasting tool: Issues and analysis. *International Journal of Forecasting* **15**: 353–375.

Winters, P. R. 1960. Forecasting sales by exponentially weighted moving average. *Management Science* **6**: 324–342.

3

Matching Supply with Demand: Aggregate Planning

3.1 Overview

In a supply chain, companies must anticipate demand (through demand forecasting) and determine how to meet it. An original equipment manufacturer (OEM) or a supplier must make decisions regarding its production levels in different time periods. Should a company maintain a constant production level, thus holding inventory during the low demand period and keeping a backlog during the high demand period? Or should a company adjust its production level based on the demand, and subcontract some of its production when the demand exceeds its production capacity? These are the type of questions that can be answered through aggregate planning.

Aggregate planning helps companies determine ideal levels of production, subcontracting, inventory, backlog, workforce, and even pricing over a specified time horizon. It has an intermediate planning time frame of between several months to a couple of years, with monthly or quarterly updates. The goal is to satisfy demand using existing production facilities while maximizing profit. Aggregate planning, as its name suggests, deals with aggregate decisions rather than stock-keeping unit (SKU) level decisions. It determines the total production level in a plant for a given time period, but does not determine the quantity of each individual SKU that will be produced. For example, if a plant produces three different sizes (32, 64, and 96 gallons, respectively) of recycle containers, an aggregate plan only determines the total quantity of containers to be produced in a given month. It does not determine how many 96-gallon containers to be produced. SKU-level production decisions are made by disaggregating an aggregate plan and then creating a master production schedule (Sipper and Bulfin 1997).

To be effective, aggregate planning must be based on accurate demand forecast. For an upstream supply chain company such as an OEM or a supplier, collaboration with downstream partners such as retailers and wholesalers could provide better demand forecasts. Some constraints for aggregate planning also come from outside supply chain partners. For example, will a company be able to subcontract part of its production? If yes, at what cost? Therefore, a company should obtain as much relevant inputs as possible from both its downstream and upstream supply chain partners in order to fully realize the potential benefits of aggregate planning. The output of aggregate planning is also of value to both downstream and upstream supply chain partners, because it defines the demand for suppliers and establishes supply constraints for customers.

Obviously, it is easier to create an aggregate plan when the demand is constant. There are several approaches to make the demand constant. The first approach is to shift demand from peak periods to nonpeak periods. This can be done through advertising or incentive programs. An example is to offer rebates during nonpeak periods to attract more customers. The second approach is to produce several products with offsetting demand patterns. These products should require similar manufacturing technology so they can be produced in the same plant without costly production line change-overs. An example is to produce lawn movers and snow blowers. They require similar assembly technology with offsetting peak demand periods. However, these approaches may not always be successful. We will need to look into elements of aggregate planning and then study different strategies to deal with fluctuating demands.

3.2 Elements of Aggregate Planning

As previously mentioned, the goal of aggregate planning is to satisfy demand using existing production facilities while maximizing profit. The most important element of aggregate planning is the *capacity* of a company's production facility. Capacity can be measured in different ways depending on the type of production system, but there is usually a natural measure. For example, the capacity of a car assembly plant may be the number of cars that can be produced during a certain period of time. Suppose the assembly plant can produce 100 cars per hour, operates an 8-h shift per day, and opens 25 days a month. Then, it has a monthly production capacity of 20,000 ($100 \times 8 \times 25$) cars. Alternatively we can express the production capacity as labor hours. For example, if the assembly plant has 500 workers that work the 8-h shift, its monthly production capacity is 100,000 ($500 \times 8 \times 25$) labor hours. We can also deduce that each car requires 5 labor hours to produce. In general, the capacity of a plant and the customer demand should be measured using the same unit. To satisfy demand, the capacity of a plant should exceed demand, at least over the long term. However, excess capacity is costly as it represents wasted investment. There are several ways to adjust the capacity of a plant, such as using overtime or adding a shift. In the car assembly plant example, if we change the shift from 8- to 10-h, then the monthly production capacity increases to 25,000 cars.

The second element of aggregate planning is the concept of an *aggregate product*. A plant usually produces several products; each of them requires somewhat different processes and has different production rates. An aggregate plan does not deal with this level of details; so these products are lumped together to form an aggregate product. For example, if a plant produces 32-gallon, 64-gallon, and 96-gallon recycle containers and the labor times needed are 2, 3, and 4 min per unit, respectively. Our aggregate plan will deal with an aggregate product, namely, recycle container. If the monthly demand for the three types of containers are 100,000, 300,000, and 100,000, respectively, then the demand for recycle containers is converted into the demand for labor time as $(100{,}000 \times 2 + 300{,}000 \times 3 + 100{,}000 \times 4)/60 = 25{,}000$ labor hours. We can say that we have a monthly demand of 500,000 (100,000 + 300,000 + 100,000) units of recycle containers and each container requires a labor time of 3 ($25{,}000 \times 60/500{,}000$) minutes. If the plant operates one 10-h shift and opens 25 days a month, then we would need 100 workers to meet the demand.

The third element of aggregate planning is *costs*. Broadly speaking, there are three categories of costs, namely, *production costs*, *inventory costs*, and *capacity change costs*. Production costs include materials, direct labor (regular time and overtime), and other costs (such as subcontracting costs) attributed to producing a product unit. Inventory costs include holding costs and backlog costs. Capacity change costs include hiring costs and layoff costs. Hiring a new worker not only incurs administrative costs but also training costs. Laying off workers incurs severance costs and costs associated with loss of goodwill. A company that lays off workers frequently will suffer from poor worker morale and find it difficult to hire workers.

The goal of aggregate planning is to maximize profit, which is equivalent to minimize costs. To simplify the problem, costs that are constant with respect to the aggregate planning decisions being made should be ignored. In general, a company should utilize the full capacity of its plant before subcontracting part of its production. Without subcontracting, costs that are considered in aggregate planning are (1) regular time labor costs, (2) overtime labor costs, (3) hiring costs, (4) layoff costs, (5) inventory holding costs, and (6) backlog costs. When subcontract is necessary, we need to include the analysis of subcontract costs versus in-house production costs including material costs.

The fourth element of aggregate planning is the *decision variables*, which include the production level, workforce size, and subcontract quantity (if subcontracting is desirable) in each planning period. The overtime hours required in each planning period can be calculated based on the production level and workforce size. The number of workers to be hired or laid off in each planning period can be calculated based on the workforce size. Inventory and backlog can be calculated based on the demand and the production level in each planning period. Therefore, when the values of the decision variables are known, the total costs of an aggregate plan can be calculated. The task of aggregate planning is to find the optimal values of the production level, workforce size, and subcontract quantity that minimizes the total costs.

3.3 Aggregate Planning Strategies

There are three basic aggregate planning strategies, namely, *level*, *capacity*, and *chase*. The level strategy maintains a constant production capacity. It keeps the workforce size constant with regularly scheduled production. Workers benefit from stable working conditions, and the company does not incur overtime production costs. This strategy is ideal when the demand is more or less constant. When there is significant demand fluctuation, this strategy will lead to inventory buildup in low-demand periods and backlog in high-demand periods. If inventory holding costs and backlog costs are low, this strategy is still desirable because it has minimal capacity change costs.

The capacity strategy is used when there is excess machine capacity (i.e., machines are not used 24 h a day and 7 days a week). The workforce size is kept constant but overtime is used when necessary. The number of overtime hours is varied in an attempt to synchronize production with demand. This strategy reduces inventory holding costs and backlog costs. However, it requires a flexible workforce. It should be used when inventory holding costs and backlog costs are relatively high and the costs associated with overtime production are relatively low.

The chase strategy synchronizes production with demand by varying the workforce size and machine capacity. This strategy results in minimal inventory holding costs and backlog costs. However, it is difficult to implement this strategy because varying the workforce size in short notice is problematic in practice. It can also have a negative impact on the morale of the workforce. It should be used when the inventory holding costs and backlog costs are very high and the costs to change workforce size are relatively low.

3.3.1 Level Strategy

Example 1: Bell Inc. produces laptop computers. The forecasted demand from January to June is shown in Table 3.1. Its assembly plant operates one shift and can accommodate a maximum workforce of 40 workers. Currently there are 30 workers. Each worker is paid a monthly salary of $3000 and can work a maximum of 160 h a month. Hiring an additional worker costs $1000, whereas laying off a worker costs $2000. Each laptop requires 1 labor hour to produce. The material cost for a laptop is $600. The inventory holding cost for a unit of laptop is $10 per month. The backlog cost for a laptop

TABLE 3.1

Laptop Demand Forecast from January to June at Bell Inc.

Month	January	February	March	April	May	June
Demand	5027	4783	5085	4962	4898	5115

TABLE 3.2

Summary of Aggregate Planning-Related Information

Maximum workforce size	40 workers
Current workforce size	30 workers
Opening inventory	1000 units
Required ending inventory	500 units
Maximum regular time hours	160 h/month
Labor hours required	1 h/unit
Material cost	$600 per unit
Regular time cost	$3000/worker/month
Hiring cost	$1000/worker
Layoff cost	$2000/worker
Inventory holding cost	$10/unit/month
Backlog cost	$100/unit/month

is $100 per month. The plant holds 1000 units of laptops at the beginning of January and would like to reduce the inventory to 500 units at the end of June. This information is summarized as shown in Table 3.2. We are asked to develop an aggregate plan for the company.

From Table 3.1, we can see that the demand is more or less constant. Therefore, we decide to use a level strategy. Our decision variables are the workforce size (W) and the monthly production level (P_t, $t = 1, 2, 3, 4, 5, 6$). Note that the workforce size is a constant over the 6-month planning horizon, but the production level may change from month to month. We may need to hire or layoff some workers at the beginning of the planning period, but we will not incur any hiring and layoff cost during the planning period. Therefore, we do not need to include hiring and layoff cost in our problem formulation. We intend to meet all of our demands. Thus, the material cost will remain constant over the planning horizon and is excluded in our problem formulation.

Let D_t, I_t, and B_t denote the demand, inventory, and backlog, respectively, in month t. Note that at least one of I_t and B_t must be zero in a given time period. If we have net inventory in time period t ($I_t > 0$), then the backlog must be 0 ($B_t = 0$). If we have no inventory in time period t ($I_t = 0$), then we may or may not have a backlog ($B_t \geq 0$). Our net inventory is $I_t - B_t$; if it is positive, then we have physical inventory; if it is negative, then we have a backlog; if it is zero, then we have neither inventory nor backlog. In a certain time period, the beginning net inventory plus the units produced minus the demand equals the ending net inventory. This relationship, called *inventory balance equation*, is shown as follows:

$$I_{t-1} - B_{t-1} + P_t - D_t = I_t - B_t \quad t = 1, 2, \ldots, T \tag{3.1}$$

where

T is the number of periods in the planning horizon
I_0 and B_0 are the initial on-hand inventory and outstanding backlog, respectively

Now we can tally up all the relevant costs as follows:

- *Regular time labor cost*: $3000 \times W \times 6$. This is simply the cost of paying for the workers in our workforce for 6 months.

- *Inventory holding cost*: $\sum_{t=1}^{6} 10 \times I_t$. Note that here we simply assume that the inventory we hold during time period t is I_t. This is just a general approximation and is sufficient for our purpose. If we know that we have a constant production rate and our products are shipped out only at the end of each time period, then we can more accurately calculate our inventory holding cost as $\sum_{t=1}^{6} 10 \times (I_{t-1} + P_t)/2$. In reality, products are shipped out multiple times in different quantities throughout each time period. To take this into account would make our problem unnecessarily complicated.

- *Backlog cost*: $\sum_{t=1}^{6} 100 \times B_t$.

Our objective is to minimize the total costs. In the mean time, we need to make sure that we do not violate the capacity constraint. First of all, the number of workers cannot exceed 40 ($W \leq 40$). With a workforce size of W, we have a maximum of $160 \times W$ production hours each month. To produce P_t units of laptops, we need $P_t \times 1$ h. Therefore, our excess capacity in each time period is $160 \times W - P_t \times 1$, which must be greater or equal to zero. In addition, to satisfy the ending inventory requirement, we have $I_6 = 500$.

One may recognize our aggregate planning problem as a constrained optimization problem, which can be solved using linear programming (Hanssmann and Hess 1960). We will discuss the formal problem formulation later. Here, we present a spreadsheet approach to develop an appropriate aggregate plan. First, we set up the spreadsheet as shown in Figure 3.1. The numbers in italic are manually entered. The other numbers are automatically calculated once we enter the formulae. The reason for entering the initial workforce size in cell C10 will become clear later on when we discuss the use of Microsoft Excel Solver to find a solution.

We can do a quick calculation to see how many workers we need. We sum up the demand over the 6-month period, subtract the initial inventory of 1000,

	A	B	C	D	E	F	G
1			**Decision Variable**				**Constraints**
2	Month	Demand	Production	Worker	Inventory	Backlog	Capacity
3					1000	0	
4	January	5027		30	0	4027	4800
5	February	4783		30	0	8810	4800
6	March	5085		30	0	13895	4800
7	April	4962		30	0	18857	4800
8	May	4898		30	0	23755	4800
9	June	5115		30	0	28870	4800
10		Worker	30				
11							
12	Cost						
13							
14	January			$90,000	$0	$402,700	
15	February			$90,000	$0	$881,000	
16	March			$90,000	$0	$1,389,500	
17	April			$90,000	$0	$1,885,700	
18	May			$90,000	$0	$2,375,500	
19	June			$90,000	$0	$2,887,000	
20							
21	Total Cost	$10,361,400					

Cell	Formula	Note
D4	=$C$10	Drag down to D9
E4	=IF(E3+C4-F3-B4>0,E3+C4-F3-B4,0)	Drag down to E9
F4	=IF(E3+C4-F3-B4<0,-(E3+C4-F3-B4),0)	Drag down to F9
G4	=D4*(160)-C4*1	Drag down to G9
D14	=D4*3000	Drag down to D19
E14	=E4*10	Drag down to E19
F14	=F4*100	Drag down to F19
B21	=SUM(D14:F19)+IF(C10>30,(C10-30)*1000,(30-C10)*2000)	

FIGURE 3.1
Spreadsheet setup for aggregate planning using the level strategy.

and add the ending inventory of 500. This comes up to 29,370, which is the number of laptops that we need to produce over the 6-month period. Because one worker can work a maximum of 160 h/month and a worker can produce 1 laptop per hour, we need an average of 31 workers a month ((29,370 × 1)/(6 × 160) = 30.6 ≈ 31). With 31 workers we can produce a maximum of 4960 units of laptop per month. We enter these numbers into the spreadsheet as shown in Figure 3.2. We can see that this results in an ending inventory of 890, which is more than what is desired. We can reduce the first month production by 390 to 4570. The spreadsheet will show that we have an excess capacity of 390 in January, and our ending inventory in June is 500. We have found a solution, with a total cost of $595,060. This solution is a feasible solution because it

	A	B	C	D	E	F	G
1			Decision Variable				Constraints
2	Month	Demand	Production	Worker	Inventory	Backlog	Capacity
3					1000	0	
4	January	5027	4960	31	933	0	0
5	February	4783	4960	31	1110	0	0
6	March	5085	4960	31	985	0	0
7	April	4962	4960	31	983	0	0
8	May	4898	4960	31	1045	0	0
9	June	5115	4960	31	890	0	0
10		Worker	31				
11							
12	Cost						
13							
14	January			$93,000	$9,330	$0	
15	February			$93,000	$11,100	$0	
16	March			$93,000	$9,850	$0	
17	April			$93,000	$9,830	$0	
18	May			$93,000	$10,450	$0	
19	June			$93,000	$8,900	$0	
20							
21	Total Cost	$618,460					
22							

FIGURE 3.2
Initial solution using the level strategy.

satisfies all the constraints. However, it may not be the best solution. We can use a trial-and-error method to change cells C4 to C10 and see if we can reduce the total cost while making sure that cell E9 is 500 (ending inventory requirement) and cells G4 to G9 are greater than 0 (does not violate capacity constraint).

The trial-and-error method is time consuming. We can use Microsoft Excel Solver to help us find a good solution quickly. Note that we say a "good" solution rather than an "optimal" solution because Excel Solver uses the Generalized Reduced Gradient nonlinear optimization algorithm. The final solution depends on the initial solution and the number of iterations in optimization. Fylstra et al. (1998) provided a detailed description of Excel Solver and how to best utilize it for linear, nonlinear, and integer programming. Here, we will just show how to use it to solve our aggregate planning problem. As shown in Figure 3.3, we tell the Solver to minimize our total cost (cell B21) by changing our decision variables (cells C4 to C10). Excel Solver requires the changeable cells to be adjacent. Because we want our workforce size to be constant from January to June, we choose to put the workforce size variable in cell C10, adjacent to the production variables, and then copy this number to cells D4 to D9. This way we can maintain the clarity of our spreadsheet while satisfying the requirement of Excel Solver.

In the "Subject to the Constraints:" box, we tell Excel Solver that the maximum workforce size is 40 ($C$10<=40); the decision variables must be integer ($C$4:$C$10=integer) and they cannot be negative ($C$4:$C$10>=0); the ending inventory must be 500 units ($E$9=500); and the excess capacity cannot be negative ($G$4:$G$9>=0). The Solver then finds a solution with 32 workers

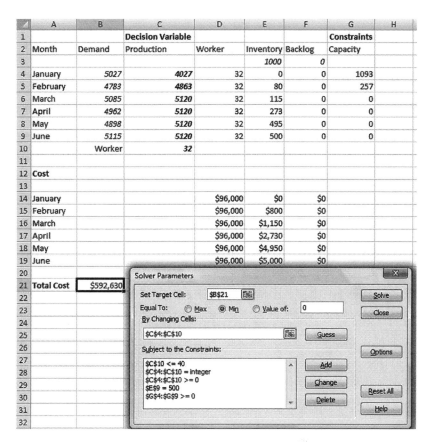

	A	B	C	D	E	F	G	H
1			**Decision Variable**				**Constraints**	
2	Month	Demand	Production	Worker	Inventory	Backlog	Capacity	
3					1000	0		
4	January	5027	4027	32	0	0	1093	
5	February	4783	4863	32	80	0	257	
6	March	5085	5120	32	115	0	0	
7	April	4962	5120	32	273	0	0	
8	May	4898	5120	32	495	0	0	
9	June	5115	5120	32	500	0	0	
10		Worker		32				
11								
12	Cost							
13								
14	January			$96,000	$0	$0		
15	February			$96,000	$800	$0		
16	March			$96,000	$1,150	$0		
17	April			$96,000	$2,730	$0		
18	May			$96,000	$4,950	$0		
19	June			$96,000	$5,000	$0		
20								
21	Total Cost	$592,630						
22								

Solver Parameters

Set Target Cell: $B$21

Equal To: ○ Max ● Min ○ Value of: 0
By Changing Cells:
$C$4:$C$10

Subject to the Constraints:
$C$10 <= 40
$C$4:$C$10 = integer
$C$4:$C$10 >= 0
$E$9 = 500
$G$4:$G$9 >= 0

[Solve] [Close] [Guess] [Options] [Add] [Change] [Reset All] [Delete] [Help]

FIGURE 3.3
Excel Solver solution using the level strategy.

as shown in Figure 3.3. The total cost is $592,630, which is $2,430 lower than our previous solution with 31 workers. This solution allows us to carry less inventory and the savings in inventory holding cost more than offset the cost in hiring an extract worker.

Again, the Solver solution depends on the initial values of the decision variables. The initial values do not necessarily need to represent a feasible solution. Generally, we can start by copying the demand into our production column and then run the Solver. Sometimes the Solver may produce a solution with fractional values that are difficult to interpret. We can manually adjust these values and run the Solver again and repeat this process until we get a satisfactory solution.

3.3.2 Capacity Strategy

Example 2: Bell Inc. recently saw some demand increase in its laptops. The demand forecast from July to December is shown in Table 3.3. At the beginning

TABLE 3.3

Laptop Demand Forecast from July to December at Bell Inc.

Month	July	August	September	October	November	December
Demand	6772	5050	6019	5244	4999	6127

of July, its assembly plant has 32 workers and an inventory of 500 laptops. The plant wants to maintain 500 units of laptop in its inventory at the end of December. If the level strategy is used, the plant needs to hire four more workers (to reach a workforce size of 36) and the total cost of the aggregate plan would be $726,010, as shown in Figure 3.4. The workers have expressed an interest of working for a maximum of 40 h of overtime per month with an hourly pay of $25. Can we reduce the cost by allowing workers to work overtime?

If we allow workers to work overtime, we need to include the overtime cost in our analysis. Let O_t denote the total overtime hours in time period t. The overtime cost is thus $\sum_{t=1}^{6} 25 \times O_t$. The total overtime hours in each time period can be calculated based on the production level and the number of workers, that is, $O_t = \max\{P_t \times 1 - 160 \times W, 0\}$. In other words, if the total hours required for production is more than the total regular time hours available, then the total overtime hours is the difference between the two; otherwise, the total overtime hours is zero. With the use of overtime, our production capacity has increased. Our excess capacity is now $(160 + 40) \times W - P_t \times 1$. Again, this excess capacity must be greater or equal to zero.

	A	B	C	D	E	F	G
1			**Decision Variable**				**Constraints**
2	Month	Demand	Production	Worker	Inventory	Backlog	Capacity
3					500	0	
4	July	6772	5760	36	0	512	0
5	August	5050	5760	36	198	0	0
6	September	6019	5760	36	0	61	0
7	October	5244	5411	36	106	0	349
8	November	4999	5760	36	867	0	0
9	December	6127	5760	36	500	0	0
10			Worker	36			
11							
12	Cost						
13							
14	July				$108,000	$0	$51,200
15	August				$108,000	$1,980	$0
16	September				$108,000	$0	$6,100
17	October				$108,000	$1,060	$0
18	November				$108,000	$8,670	$0
19	December				$108,000	$5,000	$0
20							
21	Total Cost	$726,010					

FIGURE 3.4

Level strategy solution for the new aggregate plan.

Now we can make some changes to our previous spreadsheet and use the Solver to help us find a solution. We insert an "Overtime" column and change formulae in a few cells as shown in Figure 3.5. The Solver found a solution with 33 workers and a total cost of $670,350. This solution represents a $55,660 cost saving over the level strategy.

	A	B	C	D	E	F	G	H
1			**Decision Variable**					**Constraints**
2	Month	Demand	Production	Worker	Overtime	Inventory	Backlog	Capacity
3						500	0	
4	July	6772	6272	33	992	0	0	328
5	August	5050	5244	33	0	194	0	1356
6	September	6019	5825	33	545	0	0	775
7	October	5244	5244	33	0	0	0	1356
8	November	4999	5245	33	0	246	0	1355
9	December	6127	6381	33	1101	500	0	219
10		Worker	33					
11								
12	Cost							
13								
14	July			$99,000	$24,800	$0	$0	
15	August			$99,000	$0	$1,940	$0	
16	September			$99,000	$13,625	$0	$0	
17	October			$99,000	$0	$0	$0	
18	November			$99,000	$0	$2,460	$0	
19	December			$99,000	$27,525	$5,000	$0	
20								
21	Total Cost	$670,350						

Solver Parameters

Set Target Cell: $B$21

Equal To: ○ Max ● Min ○ Value of: 0

By Changing Cells:

$C$4:$C$10

Subject to the Constraints:

$C$10 <= 40
$C$4:$C$10 = integer
$C$4:$C$10 >= 0
$F$9 = 500
$H$4:$H$9 >= 0

Solve | Close | Guess | Options | Add | Change | Reset All | Delete | Help

Cell	Formula	Note
D4	=$C$10	Drag down to D9
E4	=MAX(C4*1-160*D4,0)	Drag down to E9
F4	=IF(F3+C4-G3-B4>0,F3+C4-G3-B4,0)	Drag down to F9
G4	=IF(F3+C4-G3-B4<0,-(F3+C4-G3-B4),0)	Drag down to G9
H4	=D4*(160+40)-C4*1	Drag down to H9
D14	=D4*3000	Drag down to D19
E14	=E4*25	Drag down to E19
F14	=F4*10	Drag down to F19
G14	=G4*100	Drag down to G19
B21	=SUM(D14:G19)+IF(C10>32,(C10-32)*1000,(32-C10)*2000)	

FIGURE 3.5
Excel Solver solution using the capacity strategy.

3.3.3 Chase Strategy

Example 3: The business at Bell Inc. continues to pick up. At the end of the year, we have a new 6-month forecast as shown in Table 3.4. The workers are now demanding to be paid $50 an hour for overtime works. We still want to maintain an inventory of 500 laptops at the end of June. We updated the spreadsheet shown in Figure 3.5 by changing cells E14 to E19 (multiplying E4 to E9 by 50 instead of 25) and cell B21 (changing 32 to 33 because we have 33 workers at the end of the year). Using Excel Solver we found a solution with 40 workers and a total cost of $836,600, as shown in Figure 3.6. Can we save money by changing the workforce size?

To allow consideration of a changing workforce, we can no longer use W as our decision variable. Rather, we need a new decision variable W_t, which denotes the workforce size in time period t. W_0 is the initial workforce size. By changing the workforce size, we would incur hiring and layoff costs. Let H_t and L_t denote the number of workers hired and laid off during time period t, respectively. The workforce size at time period t is simply the workforce size

TABLE 3.4

New Laptop Demand Forecast at Bell Inc.

Month	January	February	March	April	May	June
Demand	6210	7788	7466	5200	5128	6255

	A	B	C	D	E	F	G	H
1			**Decision Variable**					**Constraints**
2	Month	Demand	Production	Worker	Overtime	Inventory	Backlog	Capacity
3						500	0	
4	January	6210	6994	40	594	1284	0	1006
5	February	7788	6504	40	104	0	0	1496
6	March	7466	7466	40	1066	0	0	534
7	April	5200	5201	40	0	1	0	2799
8	May	5128	5482	40	0	355	0	2518
9	June	6255	6400	40	0	500	0	1600
10		Worker	40					
11								
12	Cost							
13								
14	January				$120,000	$29,700	$12,840	$0
15	February				$120,000	$5,200	$0	$0
16	March				$120,000	$53,300	$0	$0
17	April				$120,000	$0	$10	$0
18	May				$120,000	$0	$3,550	$0
19	June				$120,000	$0	$5,000	$0
20								
21	Total Cost	$836,600						

FIGURE 3.6
Capacity strategy solution for the new aggregate plan.

at time period $t - 1$ plus the number of workers hired minus the number of workers laid off in time period t. Mathematically, we have

$$W_t = W_{t-1} + H_t - L_t \quad t = 1, 2, \ldots, T \tag{3.2}$$

To minimize hiring and layoff costs, if there is a need to change the workforce size in a certain time period, we will either hire workers or lay off workers but will not do both. In other words, at least one of H_t and L_t must be zero. If $W_t > W_{t-1}$, then H_t is positive and L_t must be zero. On the other hand, if $W_t < W_{t-1}$, then H_t is zero and L_t is positive. Therefore, our hiring and layoff costs are $\sum_{t=1}^{6} 1000 \times H_t + \sum_{t=1}^{6} 2000 \times L_t$.

We now make additional changes to our previous spreadsheet. Specifically, we insert a "Hired" and "Laid Off" columns and change the formulae in a few cells, as shown in Figure 3.7. Note that now our decision variables are located in cells C4 to C9 and cells D4 to D9. Therefore, we need to change the specifications in the Excel Solver user interface accordingly. The Solver found a solution with a total cost of $813,300. Therefore, using a chase strategy can save us $23,300.

3.4 Linear Programming Approach to Aggregate Planning

3.4.1 Problem Formulation

As previously mentioned, aggregate planning can be formulated as a linear programming problem. There are many commercial linear programming software tools that can solve problems involving hundreds of variables quickly. Therefore, we formulate our aggregate planning problem more generally and consider the following: (1) option of subcontracting, (2) varying working hours (both regular time and overtime) in each time period, and (3) varying costs (subcontract, regular time, overtime, hiring, lay off, inventory holding, and backlog) in each time period. For the parameters of the problem, we use the following notations:

T: number of time periods in the planning horizon

t: index of time periods, $t = 1, 2, 3, \ldots, T$

k: labor hour required to produce one product unit

D_t: forecasted number of units demanded in time period t

r_t: number of regular time hours available in time period t

v_t: number of overtime hours available in time period t

C_t^P: material cost for producing one unit of product in time period t

	A	B	C	D	E	F	G	H	I	J
1			Decision Variable							Constraints
2	Month	Demand	Production	Worker	Hired	Laid off	Overtime	Inventory	Backlog	Capacity
3				33				500	0	
4	January	6210	6399	40	7	0	0	689	0	1601
5	February	7788	7099	40	0	0	699	0	0	901
6	March	7466	7467	40	0	0	1067	1	0	533
7	April	5200	5414	34	0	6	0	215	0	1386
8	May	5128	5408	34	0	0	0	495	0	1392
9	June	6255	6260	39	5	0	20	500	0	1540
10										
11										
12	Cost									
13										
14	January			$120,000	$7,000	$0	$0	$6,890	$0	
15	February			$120,000	$0	$0	$34,950	$0	$0	
16	March			$120,000	$0	$0	$53,350	$10	$0	
17	April			$102,000	$0	$12,000	$0	$2,150	$0	
18	May			$102,000	$0	$0	$0	$4,950	$0	
19	June			$117,000	$5,000	$0	$1,000	$5,000	$0	
20				$681,000	$12,000	$12,000	$89,300	$19,000	$0	
21	Total Cost	$813,300								

Solver Parameters dialog box:

Set Target Cell: $B$21
Equal To: ○ Max ● Min ○ Value of: 0
By Changing Cells:
$C$4:$D$9

Subject to the Constraints:
$C$4:$D$9 = integer
$C$4:$D$9 >= 0
$D$4:$D$9 <= 40
$H$9 = 500
$J$4:$J$9 >= 0

[Solve] [Close] [Guess] [Options] [Add] [Change] [Reset All] [Delete] [Help]

Cell	Formula	Note
E4	=IF(D4>D3,D4-D3,0)	Drag down to E9
F4	=IF(D4<D3,D3-D4,0)	Drag down to F9
G4	=MAX(C4*1-160*D4,0)	Drag down to G9
H4	=IF(H3+C4-I3-B4>0,H3+C4-I3-B4,0)	Drag down to H9
I4	=IF(H3+C4-I3-B4<0,-(H3+C4-I3-B4),0)	Drag down to I9
J4	=D4*(160+40)-C4*1	Drag down to J9
D14	=D4*3000	Drag down to D19
E14	=E4*1000	Drag down to E19
F14	=F4*2000	Drag down to F19
G14	=E4*50	Drag down to G19
H14	=F4*10	Drag down to H19
I14	=G4*100	Drag down to I19
B21	=SUM(D14:I19)	

FIGURE 3.7
Excel Solver solution using the chase strategy.

C_t^S: cost to subcontract one unit of product in time period t

C_t^W: regular time cost of one worker in time period t

C_t^O: hourly overtime cost in time period t

C_t^H: cost to hire one worker in time period t

C_t^L: cost to lay off one worker in time period t

C_t^I: cost to hold one unit of product in inventory for time period t

C_t^B: cost to backlog one unit of product for time period t

The decision variables are listed as follows:

P_t: number of units produced in regular time in time period t

O_t: number of units produced in overtime in time period t

S_t: number of units to be subcontracted in time period t

W_t: number of workers available in time period t

H_t: number of workers hired in time period t

L_t: number of workers laid off in time period t

I_t: number of units held in inventory at the end of time period t

B_t: number of units backlogged at the end of time period t

The linear programming model is formulated as follows:

$$\text{Minimize} \quad \sum_{t=1}^{T} C_t^W W_t \qquad \text{(regular time labor cost)}$$

$$+ \sum_{t=1}^{T} k C_t^O O_t \qquad \text{(overtime labor cost)}$$

$$+ \sum_{t=1}^{T} C_t^P (P_t + O_t) \qquad \text{(material cost)}$$

$$+ \sum_{t=1}^{T} C_t^H H_t \qquad \text{(hiring cost)}$$

$$+ \sum_{t=1}^{T} C_t^L L_t \qquad \text{(lay off cost)}$$

$$+ \sum_{t=1}^{T} C_t^I I_t \quad \text{(inventory holding cost)}$$

$$+ \sum_{t=1}^{T} C_t^B B_t \quad \text{(backlog cost)}$$

$$+ \sum_{t=1}^{T} C_t^S S_t \quad \text{(subcontract cost)} \tag{3.3}$$

Subject to

$$\frac{r_t}{k} W_t - P_t \geq 0 \quad t = 1, 2, \ldots, T \quad \text{(regular time capacity constraint)} \tag{3.4}$$

$$\frac{v_t}{k} W_t - O_t \geq 0 \quad t = 1, 2, \ldots, T \quad \text{(overtime capacity constraint)} \tag{3.5}$$

$$I_{t-1} + P_t + O_t + S_t - I_t - B_{t-1} + B_t = D_t \quad t = 1, 2, \ldots, T \quad \text{(inventory balance)} \tag{3.6}$$

$$W_t - W_{t-1} - H_t + L_t = 0 \quad t = 1, 2, \ldots, T \quad \text{(workforce balance)} \tag{3.7}$$

$$P_t, S_t, W_t, H_t, L_t, I_t, B_t \geq 0 \quad t = 1, 2, \ldots, T \tag{3.8}$$

$$B_t = 0 \quad \text{(no backlog at the end of the planning period)} \tag{3.9}$$

Constraint (3.9) ensures that the total demand during the planning horizon is satisfied, which is generally the goal of aggregate planning. This will avoid the situation where a solution is found with the demand backlogged and workers laid off because it results in minimum cost. Additional constraints can be added, such as the maximum workforce size, the maximum number of workers that can be hired or laid off, the maximum inventory level, and the ending inventory level. Note that here we did not impose a constraint that the decision variables must be integer. We can certainly add such a constraint if desired. Nonetheless, it is not a big problem to have fractional unit counts in aggregate planning because we use aggregated product unit, which is fictitious. If we can hire part-time workers, then fractional worker counts also do not matter. Otherwise, we can round the solutions to obtain satisfactory results.

3.4.2 Gurobi Optimizer

Gurobi Optimizer is a software tool that can be used to solve aggregate planning problems based on a linear programming approach. It is a product of Gurobi Optimization (http://www.gurobi.com), which offers completely free, 1 year (renewable), single-user licenses to faculty, students, and staff at degree-granting academic institutions. One needs to register for a Gurobi account first. After login to the Gurobi account, one can go to Download→Gurobi Optimizer page to download the software tool. Follow the instructions in the README file to install Gurobi Optimizer. Then, go to Download→Licenses page to request a free academic license. A license code will be created in the following format: xxxxxxxx-xxxx-xxxx-xxxx-xxxxxxxxxxxx. To install this license, type "grbgetkey" followed by a space and the license code to the command/terminal prompt or the Start menu in Microsoft Windows. During this process, make sure that the computer is connected to the Internet from a recognized academic domain (e.g., any .edu address).

The linear programming model format required by Gurobi Optimizer is easy to follow. It contains four sections, summarized as follows:

- *Objective Section*: This is the first section, which begins with a header (*minimize, maximize, minimum, maximum, min,* or *max*), followed by a linear or quadratic expression that captures the objective function. The objective function contains terms that are separated by the + or − operators. A term can contain a coefficient and a variable (e.g., 160 W, note that there must be a space between the coefficient and the variable) or just a variable (e.g., P1). Note that constants (numerical values) are not recognized in the objective function. For example, if 160 is entered as a term in the objective function, it will be treated as a variable. In other words, 160 is not viewed as a numerical value but a symbol that represents a variable. Quadratic terms in the objective function must be enclosed by square brackets followed by/2, for example, $[3 x^2 + 2 x * y]/2$. The objective function can be spread over many lines.

- *Constraints Section*: This is the second section, which also begins with a header (*subject to, such that, st,* or *s.t.*) and is followed by an arbitrary number of constraints. Each constraint contains an expression, a comparison operator (=, <=, >=, or >), and a numerical value, followed by a line break. The expression are linear or quadratic terms that are separated by the + or − operators. Quadratic terms must be enclosed by square brackets but not followed by/2, for example, 5y + $[3x^2 + 2x * y]$<100.

- *Bounds Section*: This is the third section, which begins with the header *bounds* followed by an arbitrary number of lines. Each line specifies the lower bound, the upper bound, or both for a single variable (e.g., x1>5, x2<100, 0<=x3<=1). The keywords *inf* or *infinity* are used to specify infinite bounds, for example, −inf<=y<=−10. If a variable is unbounded, then the keyword *free* is used, for example, x free. It is not necessary to specify the bound for a variable if it has a lower bound of 0 and an infinite upper bound. In other words, >=0 is the default bound for variables.

- *Variable Type Section*: This is the fourth section and is optional. By default, variables are assumed to be continuous. However, they can be designated as being either binary (*binary, binaries, bin*), general integer (*general, generals, gen*), or semicontinuous (*semicontinuous, semis, semi*). In these cases, the designation is applied by first providing the appropriate header (on its own line) and then listing the variables that have the associated type.

The current version of Gurobi Optimizer is 5.0.1. In addition to the Gurobi Optimizer shell, it provides interfaces with different programming environments including Python, Java, C/C++, Visual Basic/VB.NET, R, and MATLAB®. The Python interface is probably the easiest to use. The Python program provided by Gurobi Optimizer to read and solve a linear programming model is as follows (Gurobi Optimization 2012):

```
#!/usr/bin/python

# Copyright 2012, Gurobi Optimization, Inc.

# This example reads an LP model from a file and solves it.
# If the model is infeasible or unbounded, the example turns off
# presolve and solves the model again. If the model is infeasible,
# the example computes an Irreducible Infeasible Subsystem (IIS),
# and writes it to a file

import sys
from gurobipy import *

if len(sys.argv) < 2:
    print 'Usage: lp.py filename'
    quit()

# Read and solve model

model = read(sys.argv[1])
model.optimize()

if model.status = = GRB.status.INF_OR_UNBD:
    # Turn presolve off to determine whether model is infeasible
    # or unbounded
    model.setParam(GRB.param.presolve, 0)
    model.optimize()

if model.status = = GRB.status.OPTIMAL:
    print 'Optimal objective:', model.objVal
    model.write('model.sol')
    exit(0)

elif model.status ! = GRB.status.INFEASIBLE:
    print 'Optimization was stopped with status', model.status
    exit(0)
```

```
(continued)

# Model lis infeasible - compute an Irreducible Infeasible Subsystem (IIS)

print
print "Model is infeasible"
model.computeIIS()
model.write("model.ilp")
print "IIS written to file 'model.ilp'"
```

Save this program (text file) using file name "lp.py" under the same directory as the Python application ("python.exe") installed as part of the Gurobi Optimizer software tool. If Gurobi Optimizer 5.0.1 for 64-bit Windows is installed using the default options, then this directory is c:\gurobi501\win64\python27\bin. To read and solve a linear programming model, simply save the model under the same directory (say using the file name "model_name.lp") and then open the command prompt, go to the directory, type "python lp.py model_name.lp", and hit the Enter key. The solution will be stored in a file named "model.sol".

3.4.3 Solving Aggregate Planning Problems Using Gurobi Optimizer

Now we will show how to solve aggregate planning problems using Gurobi Optimizer for the three examples discussed in Section 3.3. In Example 1, we use a level strategy and thus only require the following variables:

W: number of workers during the 6-month planning horizon

P_t: number of units produced in each month, $t = 1, 2, 3, 4, 5, 6$

I_t: number of units held in inventory at the end of each month, $t = 1, 2, 3, 4, 5, 6$

B_t: number of units backlogged at the end of each month, $t = 1, 2, 3, 4, 5, 6$

The total cost to be minimized is the sum of labor cost $\left(\sum_{t=1}^{T} C_t^W W_t = \sum_{t=1}^{6} 3,000 W_t = 3,000 \times 6 \times W = 18,000 W, \right.$ which is written as the linear term 18000 W $\Big)$, inventory holding cost $\left(\sum_{t=1}^{T} C_t^I I_t = \sum_{t=1}^{6} 10 I_t, \right.$ which needs to be written as six linear terms 10 I1 + 10 I2 + 10 I3 + 10 I4 + 10 I5 + 10 I6 $\Big)$, and backlog cost $\left(\sum_{t=1}^{T} C_t^B B_t = \sum_{t=1}^{6} 100 \times B_t, \right.$ which also needs to be written as 100 B1 + 100 B2 + 100 B3 + 100 B4 + 100 B5 + 100 B6 $\Big)$. Therefore, the objective section of the model file is written as follows:

Minimize

$$18000 \ W + 10 \ I1 + 10 \ I2 + 10 \ I3 + 10 \ I4 + 10 \ I5 + 10 \ I6 + 100 \ B1$$
$$+ 100 \ B2 + 100 \ B3 + 100 \ B4 + 100 \ B5 + 100 \ B6$$

When using the level strategy, we have two types of constraints, namely, regular time capacity constraint and inventory balance constraint. For regular time capacity constraint $((r_t/k)W_t - P_t \geq 0)$, we have $r_t = 160$, $k = 1$, $W_t = W$, and $t = 1, 2, 3, 4, 5, 6$. Therefore, we need to write six constraints. For inventory balance constraint $(I_{t-1} + P_t + O_t + S_t - I_t - B_{t-1} + B_t = D_t)$, we omit O_t and S_t because we do not use overtime and subcontract under the level strategy. Again, we need to write six constraints, one for each value of t ($t = 1, 2, 3, 4, 5, 6$). Note that D_t is known and is to the right of the comparison operator. Therefore, it does not need to be treated as a variable. When $t = 1$, the constraint is $I_0 + P_1 - I_1 - B_0 + B_1 = 5027$. It is known that $I_0 = 1000$ and $B_0 = 0$. However, we cannot write the constraint as $1000 + P1 - I1 - 0 + B1 = 5027$ because Gurobi Optimizer will treat "1000" and "0" as symbols that represent variables. We have two choices. The first choice is to write the constraint as $P1 - I1 + B1 = 4027$. The second choice is to write the constraint as $I0 + P1 - I1 - B0 + B1 = 5027$, where I0 and B0 are treated as variables. We then need to add two more constraints, namely, $I0 = 1000$ and $B0 = 0$. We will use the second choice here.

In addition to regular time capacity constraint and inventory balance constraint, we are required to hold an ending inventory of 500, that is, $I6 = 500$. Furthermore, we should make sure that there are no backlog at the end of the planning period by adding a constraint $B6 = 0$. Finally, we need to make sure that the workforce size is no more than 40. The constraints section of the model file is written as follows:

Subject To

    160 W − P1 >= 0
    160 W − P2 >= 0
    160 W − P3 >= 0
    160 W − P4 >= 0
    160 W − P5 >= 0
    160 W − P6 >= 0
    I0 + P1 − I1 − B0 + B1 = 5027
    I1 + P2 − I2 − B1 + B2 = 4783
    I2 + P3 − I3 − B2 + B3 = 5085
    I3 + P4 − I4 − B3 + B4 = 4962
    I4 + P5 − I5 − B4 + B5 = 4898
    I5 + P6 − I6 − B5 + B6 = 5115
    I0 = 1000
    B0 = 0
    I6 = 500
    B6 = 0
    W <= 40

Because all the variables have a lower bound of 0 and no upper bound, it is not necessary to specify bounds for these variables. Finally, we want all the variables to be integers, so we specify them as *general*. We can add comments (after backslash) in the model file to make it more understandable. Our final model file (which we save as "level.lp" under the same directory as "python.exe" and "lp.py") is as follows:

```
Minimize

\minimize the sum of labor cost, inventory holding cost, and backlog cost

   18000 W \regular time labor cost
   + 10 I1 + 10 I2 + 10 I3 + 10 I4 + 10 I5 + 10 I6 \inventory holding cost
   + 100 B1 + 100 B2 + 100 B3 + 100 B4 + 100 B5 + 100 B6 \backlog cost

Subject To

\regular time capacity constraint
160 W - P1 >= 0
160 W - P2 >= 0
160 W - P3 >= 0
160 W - P4 >= 0
160 W - P5 >= 0
160 W - P6 >= 0

\inventory balance constraint
I0 + P1 - I1 - B0 + B1 = 5027
I1 + P2 - I2 - B1 + B2 = 4783
I2 + P3 - I3 - B2 + B3 = 5085
I3 + P4 - I4 - B3 + B4 = 4962
I4 + P5 - I5 - B4 + B5 = 4898
I5 + P6 - I6 - B5 + B6 = 5115

\initial inventory and backlog
I0 = 1000
B0 = 0

\ending inventory and backlog
I6 = 500
B6 = 0

\workforce size limit
W <= 40

Bounds

\default is >= 0

General

\variables with known values do not need to be specified
W P1 P2 P3 P4 P5 P6 I1 I2 I3 I4 I5 B1 B2 B3 B4 B5

End
```

Now, we open the command prompt (find it under Accessories or type "cmd" to the Start menu in Microsoft Windows) and go to the directory where all of these files are saved. We then type "python lp.py level.lp" to read and solve the model as shown in Figure 3.8. We can see that an optimal solution is found,

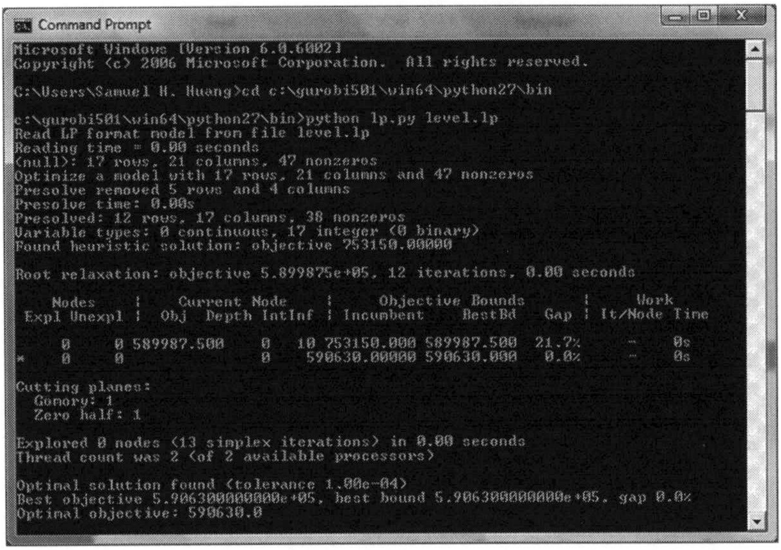

```
Microsoft Windows [Version 6.0.6002]
Copyright (c) 2006 Microsoft Corporation.  All rights reserved.

C:\Users\Samuel H. Huang>cd c:\gurobi501\win64\python27\bin

c:\gurobi501\win64\python27\bin>python lp.py level.lp
Read LP format model from file level.lp
Reading time = 0.00 seconds
(null): 17 rows, 21 columns, 47 nonzeros
Optimize a model with 17 rows, 21 columns and 47 nonzeros
Presolve removed 5 rows and 4 columns
Presolve time: 0.00s
Presolved: 12 rows, 17 columns, 38 nonzeros
Variable types: 0 continuous, 17 integer (0 binary)
Found heuristic solution: objective 753150.00000

Root relaxation: objective 5.899875e+05, 12 iterations, 0.00 seconds

    Nodes    |    Current Node    |     Objective Bounds      |     Work
 Expl Unexpl |  Obj  Depth IntInf | Incumbent    BestBd   Gap | It/Node Time

     0     0 589987.500    0   10 753150.000 589987.500  21.7%     -    0s
*    0     0               0    590630.00000 590630.000   0.0%     -    0s

Cutting planes:
  Gomory: 1
  Zero half: 1

Explored 0 nodes (13 simplex iterations) in 0.00 seconds
Thread count was 2 (of 2 available processors)

Optimal solution found (tolerance 1.00e-04)
Best objective 5.906300000000e+05, best bound 5.906300000000e+05, gap 0.0%
Optimal objective: 590630.0
```

FIGURE 3.8
Running Gurobi Optimizer under Microsoft Windows command prompt.

where the minimum cost is $590,630. The complete solution is stored in the file "model.sol" under the same directory. The content of this file is as follows:

```
# Objective value = 590630
W 32
I1 0
I2 80
I3 115
I4 273
I5 495
I6 500
B1 0
B2 0
B3 0
B4 0
B5 0
B6 0
P1 4027
P2 4863
P3 5120
P4 5120
P5 5120
P6 5120
I0 1000
B0 0
```

Note that this solution is identical to the one we found using Excel Solver (see Figure 3.3), except that the total cost is $2000 less. This is because we have not added the hiring cost at the beginning of the planning period. Because the initial workforce is 30 workers and the optimal solution requires 32 workers, we need to hire 2 workers at the beginning of the planning period, which costs $2000.

In Example 2, we use a capacity strategy that requires the use of overtime. Therefore, we need to add variables O_t ($t = 1, 2, 3, 4, 5, 6$), which is the number of product units produced in overtime in each month. We need to add overtime labor cost $\left(\sum_{t=1}^{T} kC_t^O O_t = \sum_{t=1}^{6} 1 \times 25 \times O_t \right)$ to the objective function, which needs to be written as 25 O1 + 25 O2 + 25 O3 + 25 O4 + 25 O5 + 25 O6. We also need to add overtime capacity constraint $((v_t/k)W_t - O_t \geq 0)$. We have $v_t = 40$, $k = 1$, $W_t = W$, and $t = 1, 2, 3, 4, 5, 6$; thus we need to write six constraints. The demand in each month is different, so the inventory balance constraints need to be updated. The initial inventory also needs to be updated. Finally, we need to specify that the variables O_t are general integers. Now we have a new model file (which we save as "capacity.lp") as follows:

```
Minimize

\minimize the sum of labor cost, inventory holding cost, and backlog cost

   18000 W \regular time labor cost
   + 25 O1 + 25 O2 + 25 O3 + 25 O4 + 25 O5 + 25 O6 \overtime labor cost
   + 10 I1 + 10 I2 + 10 I3 + 10 I4 + 10 I5 + 10 I6 \inventory holding cost
   + 100 B1 + 100 B2 + 100 B3 + 100 B4 + 100 B5 + 100 B6 \backlog cost

Subject To

\regular time capacity constraint
   160 W - P1 >= 0
   160 W - P2 >= 0
   160 W - P3 >= 0
   160 W - P4 >= 0
   160 W - P5 >= 0
   160 W - P6 >= 0

\overtime capacity constraint
   40 W - O1 >= 0
   40 W - O2 >= 0
   40 W - O3 >= 0
   40 W - O4 >= 0
   40 W - O5 >= 0
   40 W - O6 >= 0

\inventory balance constraint
   I0 + P1 + O1 - I1 - B0 + B1 = 6772
   I1 + P2 + O2 - I2 - B1 + B2 = 5050
   I2 + P3 + O3 - I3 - B2 + B3 = 6019
   I3 + P4 + O4 - I4 - B3 + B4 = 5244
   I4 + P5 + O5 - I5 - B4 + B5 = 4999
   I5 + P6 + O6 - I6 - B5 + B6 = 6127
```

(continued)

(continued)

```
\initial inventory and backlog
I0 = 500
B0 = 0

\ending inventory and backlog
I6 = 500
B6 = 0

\workforce size limit
W >= 40

Bounds

\default is >= 0

General

\variables with known values do not need to be specified
W P1 P2 P3 P4 P5 P6 O1 O2 O3 O4 O5 O6 I1 I2 I3 I4 I5 B1 B2 B3 B4 B5

End
```

We solve this model by typing "python lp.py capacity.lp" in the command prompt and obtain the solution as follows:

```
# Objective value = 668105
W 33
O1 992
O2 0
O3 509
O4 0
O5 0
O6 1030
I1 0
I2 230
I3 0
I4 36
I5 317
I6 500
B1 0
B2 0
B3 0
B4 0
B5 0
B6 0
P1 5280
P2 5280
P3 5280
P4 5280
P5 5280
P6 5280
I0 500
B0 0
```

This solution is slightly different from the one we obtained using Excel Solver (see Figure 3.5). Taking into account the additional worker (the solution calls for 33 workers and we have 32 workers initially) that we need to hire at the beginning of the planning period, the total cost for the solution obtained using Gurobi Optimizer is $669,105. This is $1245 less than that obtained using Excel Solver solution, which indicates that Gurobi Optimizer performs better than Excel Solver. However, Excel has a better user interface. Using Excel, one can easily set up a spreadsheet that clearly shows details of the solution, along with different types of costs under different planning periods. To take advantage of Gurobi Optimizer's superior optimization performance while utilizing Excel's better presentation, we can simply enter Gurobi Optimizer solution to the Excel spreadsheet that we previously set up. Refer to Figure 3.5, we only need to change the decision variables in column C. The number of workers remains at 33 (cell C10). The number of units produced in each month is the sum of units produced in regular time and the units produced in overtime, that is, $P_t + O_t$. In the first month, we have P1 = 5280 and O1 = 992; thus we need to produce 5280 + 992 = 6272 units of laptops. We enter 6272 in cell C4 of the spreadsheet that we previously set up. Similarly, we can calculate the production units for the rest of the months and enter them in cells C5 to C9. Now we have a spreadsheet solution as shown in Figure 3.9, which has a lower cost than the one shown in Figure 3.5. Note that in the EXCEL spreadsheet that we previously setup,

	A	B	C	D	E	F	G	H
1			**Decision Variable**					**Constraints**
2	Month	Demand	Production	Worker	Overtime	Inventory	Backlog	Capacity
3						500	0	
4	July	6772	6272	33	992	0	0	328
5	August	5050	5280	33	0	230	0	1320
6	September	6019	5789	33	509	0	0	811
7	October	5244	5280	33	0	36	0	1320
8	November	4999	5280	33	0	317	0	1320
9	December	6127	6310	33	1030	500	0	290
10			Worker	33				
11								
12	Cost							
13								
14	July				$99,000	$24,800	$0	$0
15	August				$99,000	$0	$2,300	$0
16	September				$99,000	$12,725	$0	$0
17	October				$99,000	$0	$360	$0
18	November				$99,000	$0	$3,170	$0
19	December				$99,000	$25,750	$5,000	$0
20								
21	Total Cost	$669,105						

FIGURE 3.9
Gurobi Optimizer solution in Excel spreadsheet: the capacity strategy.

the overtime is in hours, not in product units. The formula (for Cell E4) shown in Figure 3.5 took care of the conversion.

Now we look at Example 3, where the chase strategy is used. Instead of using the variable W for the size of the workforce, we need to use variables W_t ($t = 1, 2, 3, 4, 5, 6$) because the workforce size is allowed to vary in different months. We also need to add variables H_t and L_t ($t = 1, 2, 3, 4, 5, 6$) to represent the number of workers hired and laid off each month. These variables are general integers. We need to change the regular time labor cost in the objective function, that is, using $\sum_{t=1}^{6} 3000 \times W_t$ instead of $18{,}000 \times W$. We need to change hourly overtime cost to \$50 from \$25. We also need to add the hiring $\left(\sum_{t=1}^{T} C_t^H H_t = \sum_{t=1}^{6} 1000 \times H_t \right)$ and lay off $\left(\sum_{t=1}^{T} C_t^L L_t = \sum_{t=1}^{6} 2000 \times L_t \right)$ costs to the objective function. Next, we need to update regular time and overtime capacity constraints and workforce size limit by changing W to W_t in the corresponding month. We also need to update the demand of each month for the inventory balance constraints. Finally, we need to add workforce balance constraints ($W_t - W_{t-1} - H_t + L_t = 0$) and specify that the initial workforce size is 33 (W0 = 33). We now have a new model file (which we save as "chase.lp") as follows:

```
Minimize

\minimize the sum of labor cost, inventory holding cost, and backlog cost
3000 W1 + 3000 W2 + 3000 W3 + 3000 W4 + 3000 W5 + 3000 W6 \regular time labor cost
+ 50 O1 + 50 O2 + 50 O3 + 50 O4 + 50 O5 + 50 O6 \overtime labor cost
+ 10 I1 + 10 I2 + 10 I3 + 10 I4 + 10 I5 + 10 I6 \inventory holding cost
+ 100 B1 + 100 B2 + 100 B3 + 100 B4 + 100 B5 + 100 B6 \backlog cost
+ 1000 H1 + 1000 H2 + 1000 H3 + 1000 H4 + 1000 H5 + 1000 H6 \hiring cost
+ 2000 L1 + 2000 L2 + 2000 L3 + 2000 L4 + 2000 L5 + 2000 L6 \lay off cost

Subject To

\regular time capacity constraint
160 W1 - P1 >= 0
160 W2 - P2 >= 0
160 W3 - P3 >= 0
160 W4 - P4 >= 0
160 W5 - P5 >= 0
160 W6 - P6 >= 0

\overtime capacity constraint
40 W1 - O1 >= 0
40 W2 - O2 >= 0
40 W3 - O3 >= 0
40 W4 - O4 >= 0
40 W5 - O5 >= 0
40 W6 - O6 >= 0

\inventory balance
I0 + P1 + O1 - I1 - B0 + B1 = 6210
I1 + P2 + O2 - I2 - B1 + B2 = 7788
```

(continued)

```
I2 + P3 + O3 - I3 - B2 + B3 = 7466
I3 + P4 + O4 - I4 - B3 + B4 = 5200
I4 + P5 + O5 - I5 - B4 + B5 = 5128
I5 + P6 + O6 - I6 - B5 + B6 = 6255

\initial inventory and backlog
I0 = 500
B0 = 0

\ending inventory and backlog
I6 = 500
B6 = 0

\workforce size limit
W1 <= 40
W2 <= 40
W3 <= 40
W4 <= 40
W5 <= 40
W6 <= 40

\workforce balance
W1 - W0 - H1 + L1 = 0
W2 - W1 - H2 + L2 = 0
W3 - W2 - H3 + L3 = 0
W4 - W3 - H4 + L4 = 0
W5 - W4 - H5 + L5 = 0
W6 - W5 - H6 + L6 = 0

\initial workforce size
W0 = 33

Bounds

\default is >= 0

General

\variables with known values do not need to be specified
W1 W2 W3 W4 W5 W6 P1 P2 P3 P4 P5 P6 O1 O2 O3 O4 O5 O6
H1 H2 H3 H4 H5 H6 L1 L2 L3 L4 L5 L6 I1 I2 I3 I4 I5 B1 B2 B3 B4 B5

End
```

We solve this model by typing "python lp.py chase.lp" in the command prompt and obtain the solution as follows:

```
# Objective value = 812080
W1 40
W2 40
W3 40
W4 35
W5 35
W6 37
```

(continued)

(continued)

```
O1  0
O2  698
O3  1066
O4  0
O5  0
O6  0
I1  690
I2  0
I3  0
I4  363
I5  835
I6  500
B1  0
B2  0
B3  0
B4  0
B5  0
B6  0
H1  7
H2  0
H3  0
H4  0
H5  0
H6  2
L1  0
L2  0
L3  0
L4  5
L5  0
L6  0
P1  6400
P2  6400
P3  6400
P4  5563
P5  5600
P6  5920
I0  500
B0  0
W0  33
```

Again, this is a slightly better solution than the one we obtained using Excel Solver (see Figure 3.7). We enter this solution to the Excel spreadsheet to take advantage of the better presentation format as shown in Figure 3.10.

	A	B	C	D	E	F	G	H	I	J
1			Decision Variable							Constraints
2	Month	Demand	Production	Worker	Hired	Laid off	Overtime	Inventory	Backlog	Capacity
3				33				500	0	
4	January	6210	6400	40	7	0	0	690	0	1600
5	February	7788	7098	40	0	0	698	0	0	902
6	March	7466	7466	40	0	0	1066	0	0	534
7	April	5200	5563	35	0	5	0	363	0	1437
8	May	5128	5600	35	0	0	0	835	0	1400
9	June	6255	5920	37	2	0	0	500	0	1480
10										
11										
12	Cost									
13										
14	January			$120,000	$7,000	$0	$0	$6,900	$0	
15	February			$120,000	$0	$0	$34,900	$0	$0	
16	March			$120,000	$0	$0	$53,300	$0	$0	
17	April			$105,000	$0	$10,000	$0	$3,630	$0	
18	May			$105,000	$0	$0	$0	$8,350	$0	
19	June			$111,000	$2,000	$0	$0	$5,000	$0	
20										
21	Total Cost	$812,080								

FIGURE 3.10
Gurobi Optimizer solution in Excel spreadsheet: the chase strategy.

3.5 Case Studies

Case Study 3.5.1

The Container Pro Company produces 32-gallon, 64-gallon, and 96-gallon recycle containers. The sales price, material cost, inventory holding cost, backlog cost, labor time, and subcontract cost are summarized in Table 3.5. The recycle containers are produced using molding machines. The company has a floor space that can hold 100 molding machines but currently has only 80 machines. Each machine requires one worker to operate. There are 80 workers who work an 8-h shift

TABLE 3.5

Summary of Recycle Container-Related Information

	32-Gallon	64-Gallon	96-Gallon
Sales price (per unit)	$10	$15	$20
Material cost (per unit)	$5	$7	$10
Inventory holding cost (per unit per month)	$0.2	$0.3	$0.4
Backlog cost (per unit per month)	$1	$1.5	$2
Labor time (per unit)	2 min	3 min	4 min
Subcontract cost (per unit)	$8	$12	$16

TABLE 3.6

Container Demand Forecast and Working Days

Month	Recycle Container Demand			Working Days
	32-Gallon	64-Gallon	96-Gallon	
January	100,000	120,000	80,000	21
February	100,000	90,000	100,000	18
March	120,000	110,000	90,000	20
April	110,000	150,000	110,000	20
May	95,000	135,000	120,000	21
June	115,000	100,000	200,000	20
July	150,000	110,000	190,000	20
August	125,000	95,000	150,000	22
September	110,000	120,000	110,000	20
October	90,000	110,000	120,000	22
November	100,000	125,000	100,000	17
December	96,000	100,000	85,000	19

per day. Each worker is paid $2000 per month. To hire a new worker costs $500 and to lay off an existing worker costs $800. Each worker can work a maximum of 40 h of overtime per month. The overtime hourly cost is $20. The monthly demand forecast and working days are summarized in Table 3.6. Develop an appropriate aggregate plan. The purchase price for a molding machine is $200,000 and its maintenance cost is $25,000 per year. If the useful life of a molding machine is 15 years, does it make sense for the company to invest in additional molding machines?

Because the company produces three types of recycle containers, we need to create an aggregate product and determine the sales price, material cost, inventory holding cost, backlog cost, and subcontract cost for this aggregate product. Over the 12-month planning horizon, the total demand for the three types of recycle containers are 1,311,000 units, 1,365,000 units, and 1,455,000 units, respectively. We can convert this demand into required production hours as 43,700, 68,250, and 97,000 h, respectively. For simplicity, we create an aggregate recycle container (ARC) that requires 1 h of production time. The ARC can be viewed as a composite of 6.274 units, 6.533 units, and 6.963 units of 32-gallon, 64-gallon, and 96-gallon recycle containers, respectively. Its sales price and cost information can then be calculated. The calculation is shown in Figure 3.11.

We now convert the monthly demand for the three containers into the demand for ARC and calculate the available regular time hours as shown in Figure 3.12. Refer to Section 3.4, we have 8 categories of decision variables and a 12-period time horizon. Therefore, we have a linear

	A	B	C	D	E
1		32-gallon	64-gallon	96-gallon	ARC
2	Sales price (/unit)	$10	$15	$20	$300.00
3	Material cost (/unit)	$5	$7	$10	$146.73
4	Inventory holding cost (/unit/month)	$0.20	$0.30	$0.40	$6.00
5	Backlog cost (/unit/month)	$1	$1.50	$2	$30.00
6	Subcontract cost (/unit)	$8	$12	$16	$240.00
7	Labor time (minute/unit)	2	3	4	
8	Demand	1311000	1365000	1455000	
9	Total laobr time (hour)	43700	68250	97000	
10	ARC component	6.2742283	6.532663	6.9633884	

Cell	Formula	Note
B9	= B8*B7/60	Drag right to D9
B10	= B9/SUM($B9:$D9)*60/B7	Drag right to D10
E2	= SUMPRODUCT(B2:D2,B$10:D$10)	Drag down to E6

FIGURE 3.11
Calculation of sales price and cost information for the aggregate product.

	A	B	C	D	E	F	G
1	Month	Recycle Container Demand			Working	ARC	Regular
2		32-gallon	64-gallon	96-gallon	Days	Demand	Time
3	January	100,000	120,000	80,000	21	14667	168
4	February	100,000	90,000	100,000	18	14500	144
5	March	120,000	110,000	90,000	20	15500	160
6	April	110,000	150,000	110,000	20	18500	160
7	May	95,000	135,000	120,000	21	17917	168
8	June	115,000	100,000	200,000	20	22167	160
9	July	150,000	110,000	190,000	20	23167	160
10	August	125,000	95,000	150,000	22	18917	176
11	September	110,000	120,000	110,000	20	17000	160
12	October	90,000	110,000	120,000	22	16500	176
13	November	100,000	125,000	100,000	17	16250	136
14	December	96,000	100,000	85,000	19	13867	152

Cell	Formula	Note
F3	=ROUND((B3*2+C3*3+D3*4)/60,0)	Drag down to F14
G4	=8*E3	Drag down to G14

FIGURE 3.12
Calculation of monthly demand for ARC and available regular time hours.

programming problem with 96 decision variables. For convenience sake, we restrict all the variables to integers. The linear programming model file for Gurobi Optimizer is as follows:

```
Minimize

\regular time labor cost
2000 W1 + 2000 W2 + 2000 W3 + 2000 W4 + 2000 W5 + 2000 W6
+ 2000 W7 + 2000 W8 + 2000 W9 + 2000 W10 + 2000 W11 + 2000 W12
\overtime labor cost
+ 20 O1 + 20 O2 + 20 O3 + 20 O4 + 20 O5 + 20 O6
+ 20 O7 + 20 O8 + 20 O9 + 20 O10 + 20 O11 + 20 O12
\material cost
+ 146.73 P1 + 146.73 P2 + 146.73 P3 + 146.73 P4 + 146.73 P5 + 146.73 P6
+ 146.73 P7 + 146.73 P8 + 146.73 P9 + 146.73 P10 + 146.73 P11 + 146.73 P12
+ 146.73 O1 + 146.73 O2 + 146.73 O3 + 146.73 O4 + 146.73 O5 + 146.73 O6
+ 146.73 O7 + 146.73 O8 + 146.73 O9 + 146.73 O10 + 146.73 O11 + 146.73 O12
\hiring cost
+ 500 H1 + 500 H2 + 500 H3 + 500 H4 + 500 H5 + 500 H6
+ 500 H7 + 500 H8 + 500 H9 + 500 H10 + 500 H11 + 500 H12
\lay off cost
+ 800 L1 + 800 L2 + 800 L3 + 800 L4 + 800 L5 + 800 L6
+ 800 L7 + 800 L8 + 800 L9 + 800 L10 + 800 L11 + 800 L12
\inventory holding cost
+ 6 I1 + 6 I2 + 6 I3 + 6 I4 + 6 I5 + 6 I6
+ 6 I7 + 6 I8 + 6 I9 + 6 I10 + 6 I11 + 6 I12
\backlog cost
+ 30 B1 + 30 B2 + 30 B3 + 30 B4 + 30 B5 + 30 B6
+ 30 B7 + 30 B8 + 30 B9 + 30 B10 + 30 B11 + 30 B12
\subcontract cost
+ 240 S1 + 240 S2 + 240 S3 + 240 S4 + 240 S5 + 240 S6
+ 240 S7 + 240 S8 + 240 S9 + 240 S10 + 240 S11 + 240 S12


Subject To

\regular time capacity constraint
168 W1 - P1 >= 0
144 W2 - P2 >= 0
160 W3 - P3 >= 0
160 W4 - P4 >= 0
168 W5 - P5 >= 0
160 W6 - P6 >= 0
160 W7 - P7 >= 0
176 W8 - P8 >= 0
160 W9 - P9 >= 0
176 W10 - P10 >= 0
136 W11 - P11 >= 0
152 W12 - P12 >= 0


\overtime capacity constraint
40 W1 - O1 >= 0
40 W2 - O2 >= 0
40 W3 - O3 >= 0
40 W4 - O4 >= 0
40 W5 - O5 >= 0
40 W6 - O6 >= 0
40 W7 - O7 >= 0
40 W8 - O8 >= 0
40 W9 - O9 >= 0
40 W10 - O10 >= 0
40 W11 - O11 >= 0
40 W12 - O12 >= 0


\inventory balance
I0 + P1 + O1 + S1 - I1 - B0 + B1 = 14667
I1 + P2 + O2 + S2 - I2 - B1 + B2 = 14500
I2 + P3 + O3 + S3 - I3 - B2 + B3 = 15500
I3 + P4 + O4 + S4 - I4 - B3 + B4 = 18500
I4 + P5 + O5 + S5 - I5 - B4 + B5 = 17917
I5 + P6 + O6 + S6 - I6 - B5 + B6 = 22167
I6 + P7 + O7 + S7 - I7 - B6 + B7 = 23167
I7 + P8 + O8 + S8 - I8 - B7 + B8 = 18917
I8 + P9 + O9 + S9 - I9 - B8 + B9 = 17000
```

```
(continued)

I9 + P10 + O10 + S10 - I10 - B9 + B10 = 16500
I10 + P11 + O11 + S11 - I11 - B10 + B11 = 16250
I11 + P12 + O12 + S12 - I12 - B11 + B12 = 13867

\initial inventory and backlog
I0 = 0
B0 = 0

\ending backlog
B12 = 0

\workforce balance
W1 - W0 - H1 + L1 = 0
W2 - W1 - H2 + L2 = 0
W3 - W2 - H3 + L3 = 0
W4 - W3 - H4 + L4 = 0
W5 - W4 - H5 + L5 = 0
W6 - W5 - H6 + L6 = 0
W7 - W6 - H7 + L7 = 0
W8 - W7 - H8 + L8 = 0
W9 - W8 - H9 + L9 = 0
W10 - W9 - H10 + L10 = 0
W11 - W10 - H11 + L11 = 0
W12 - W11 - H12 + L12 = 0

\initial workforce
W0 = 80

\workforce size limit
W1 <= 80
W2 <= 80
W3 <= 80
W4 <= 80
W5 <= 80
W6 <= 80
W7 <= 80
W8 <= 80
W9 <= 80
W10 <= 80
W11 <= 80
W12 <= 80

Bounds

\default is >= 0

Generals

W1 W2 W3 W4 W5 W6 W7 W8 W9 W10 W11 W12
O1 O2 O3 O4 O5 O6 O7 O8 O9 O10 O11 O12
H1 H2 H3 H4 H5 H6 H7 H8 H9 H10 H11 H12
L1 L2 L3 L4 L5 L6 L7 L8 L9 L10 L11 L12
I1 I2 I3 I4 I5 I6 I7 I8 I9 I10 I11 I12
B1 B2 B3 B4 B5 B6 B7 B8 B9 B10 B11 B12
S1 S2 S3 S4 S5 S6 S7 S8 S9 S10 S11 S12
P1 P2 P3 P4 P5 P6 P7 P8 P9 P10 P11 P12

End
```

The solution is shown in Table 3.7. The total cost for this solution is $35,022,894. A total of 208,952 ARCs would be sold, which generates revenue of $62,685,600. The gross profit is $27,662,706. We observe that the company's current production capacity is fully utilized and subcontract is necessary. Therefore, we consider purchasing another 20 molding machines to fully occupy the company floor space. This means we can have a maximum workforce size of 100. We revise our problem formulation by changing the workforce constraint to less than or equal to 100 in each month and obtain a new solution as shown in Table 3.8. The total cost for this new solution is $33,520,269. The gross profit is $29,165,331.

TABLE 3.7

Linear Programming Solution with 80 Machines

Month	Worker	Hired	Laid Off	Inventory	Backlog	Subcontract	Regular Time Production	Overtime Production
1	80	0	0	1973	0	0	13,440	3200
2	80	0	0	2193	0	0	11,520	3200
3	80	0	0	2693	0	0	12,800	3200
4	80	0	0	193	0	0	12,800	3200
5	80	0	0	0	0	1084	13,440	3200
6	80	0	0	0	0	6167	12,800	3200
7	80	0	0	0	0	7167	12,800	3200
8	80	0	0	0	0	1637	14,080	3200
9	80	0	0	0	103	897	12,800	3200
10	80	0	0	677	0	0	14,080	3200
11	80	0	0	0	1493	0	10,880	3200
12	80	0	0	0	0	0	12,160	3200

TABLE 3.8

New Linear Programming Solution with 100 Machines

Month	Worker	Hired	Laid Off	Inventory	Backlog	Subcontract	Regular Time Production	Overtime Production
1	88	8	0	117	0	0	14,784	0
2	100	12	0	17	0	0	14,400	0
3	100	0	0	951	0	0	16,000	434
4	100	0	0	2451	0	0	16,000	4000
5	100	0	0	5334	0	0	16,800	4000
6	100	0	0	3167	0	0	16,000	4000
7	100	0	0	0	0	0	16,000	4000
8	100	0	0	0	0	0	17,600	1317
9	100	0	0	0	0	0	16,000	1000
10	100	0	0	1100	0	0	17,600	0
11	100	0	0	0	0	0	13,600	1550
12	91	0	9	0	0	0	13,832	35

This is an increase of $1,502,625 over that of the aggregate plan with current production capacity. Note that each molding machine requires $25,000 for annual maintenance. With 20 additional molding machines, the maintenance cost is $500,000 per year. Therefore, we only have a net profit increase of $1,002,625 per year. Because each molding machine costs $200,000, it will take us about 4 years to recoup our initial investment, assuming a 0% discount rate and the same level of demand each year. Because the molding machine has a 15-year useful life, it makes sense to invest in additional molding machines.

Case Study 3.5.2

Kenwood Clocks is a small manufacturer of high-end grandfather clocks. Its business has been stable with an annual demand between 3000 and 3500 grandfather clocks. There are 60 workers in the company. These workers are highly skilled. Hiring and training a new worker is quite expensive and costs an average of $20,000 to do so. As such, these workers are treated very well by the company's owner, Al Kenwood. In fact, Al promised these workers that the company will try its best to keep them on payroll even if business is bad. These workers work an 8-h regular time shift and their average annual salary is $48,000. They are willing to work an additional 4-h overtime shift when necessary, with an hourly pay of $40. On average, there are 480 regular time hours and up to 240 overtime hours each quarter. It takes 40 worker hours to manufacture a grandfather clock. The inventory holding cost is $300 per clock per quarter, whereas the backlog cost is $100 per clock per quarter. Al has never attempted to develop an optimal aggregate plan. He simply asks workers to work the regular time shift and when necessary ask them to work overtime in order to satisfy demand. In the past year, the demands were 900, 720, 880, and 800 for quarters I, II, III, and IV, respectively. Al was able to satisfy all the demands without holding inventory or backlogging demands. He is quite happy with his ad hoc planning approach. Can you convince him that developing an optimal aggregate plan is beneficial?

With 60 workers working 480 regular time hours and up to 240 overtime hours per quarter, Kenwood Clocks can produce $60 \times 480/40 = 720$ grandfather clocks in regular time and $60 \times 240/40 = 360$ grandfather clocks in overtime each quarter. This production capacity is sufficient to meet customer demands. With Al's plan, 7200, 0, 6400, and 3200 h of overtime are required for quarters I, II, III, and IV, respectively. The overtime labor cost is $(7,200 + 0 + 6,400 + 3,200) \times 40 = \$672,000$. The regular time labor cost is $48,000 \times 60 = \$2,888,000$. The total labor cost is $3,552,000.

To develop an optimal aggregate plan, we use Gurobi Optimizer and obtain an aggregate plan that calls for 69 workers. This plan is presented in a spreadsheet as shown in Figure 3.13. Some demands in quarter I and III are backlogged and inventory is held in quarter II. The total cost

	A	B	C	D	E	F	G	H
1			**Decision Variable**					**Constraints**
2	Quarter	Demand	Production	Worker	Overtime	Inventory	Backlog	Capacity
3						*0*	*0*	
4	I	900	*828*	69	0	0	72	16560
5	II	720	*816*	69	0	24	0	17040
6	III	880	*828*	69	0	0	28	16560
7	IV	800	*828*	69	0	0	0	16560
8		Worker		*69*				
9								
10	Cost							
11				$828,000	$0	$0	$7,200	
12				$828,000	$0	$7,200	$0	
13				$828,000	$0	$0	$2,800	
14				$828,000	$0	$0	$0	
15								
16	Total Cost	$3,329,200						

FIGURE 3.13
Optimal aggregate plan for Kenwood Clocks.

is $3,329,200. This is $222,800 less than Al Kenwood's plan. Even taking into account the one-time cost of hiring nine additional workers (20,000 × 9 = $180,000), this optimal plan still saves the company $42,800. From this case study, we can see that it is possible to meet customer demand without making a formal aggregate plan when there is sufficient production capacity. However, because overtime capacity is required that has a higher cost than regular time capacity, it is beneficial to develop an optimal aggregate plan in order to minimize cost.

Case Study 3.5.3

You presented the result of the optimal aggregate plan from Case 3.2 to Al Kenwood and suggested that he hire six more workers. Al was not totally convinced the usefulness of formal aggregate planning through optimization. He stated that your analysis is simply "Monday morning quarterbacking." Your analysis requires knowing the exact demand in each quarter, which is impossible in reality. Al indicated that based on his experience, once in a while the quarterly demand could be as low as 600. Because he has made a promise not to lay off workers, he is reluctant to increase the workforce size. He also indicated that quarterly demand in the past has never exceeded 1050, which can be met with the current workforce. However, Al conceded that reducing overtime hours saves the company money. Because yearly demand has never been lower than 3000, he would consider hiring two or three more workers but not more. Is Al correct?

Al Kenwood has a point. Aggregate plans are made based on forecasted demands, which are always inaccurate. An optimal aggregate plan on paper may not be optimal in real world when actual demands deviate considerably from the forecasted demands. Therefore, it is beneficial to conduct what-if analysis when developing an aggregate plan. Al is most concerned about low-demand quarters. Because the yearly demand is unlikely to be lower than 3000 and the highest quarterly demand does not exceed 1050, we will not have more than two quarters of lowest demand of 600 (if we have three lowest demand quarters, then the demand in the other quarter would be at least 1200, which is quite impossible based on Al's experience). Therefore, a worst-case scenario is that our demand is 600, 600, 750, and 1050 for quarters I, II, III, and IV. Under these circumstances, the optimal aggregate plan calls for 62 workers. Therefore, if Al believes that future demands will be low, then his idea of hiring only two to three more workers is justified.

However, Al Kenwood may be too conservative. We should also consider other demand scenarios. The best-case scenario is that we have a high yearly demand of 3500 grandfather clocks and it is evenly distributed in each quarter (875 per quarter). Under this scenario, the optimal aggregate plan calls for 73 workers. Therefore, we have a lower bound of 62 workers and an upper bound of 73 workers. To determine how many more workers to hire, we should evaluate the probability of each scenario. Suppose Al agreed to consider alternative scenarios and estimated that there are 30% chance of the worst-case scenario and 10% chance of the best-case scenarios previously described, and 60% of an average scenario where the demands are 800, 825, 800, and 825 in quarters I, II, III, and IV, respectively. We can then evaluate the expected cost of having between 62 and 73 workers. Table 3.9 shows the optimal aggregate plan costs for having 62, 67, and 73 workers under different scenarios. We can see that 62 workers are ideal for the low-demand scenario. However, if demand increases, the aggregate plan cost increases substantially because of increased use of overtime hours. The expected cost of having 62 workers is 3,228,600 × 30% + 3,414,400 × 60% + 3,814,400 × 10% = $3,398,660. On the other hand, with 73 workers, the aggregate plan cost does not change much under different demand scenarios because virtually no overtime is required.

TABLE 3.9

Aggregate Planning Costs under Different Scenarios

	Scenario			
Workers	Low Demand (30%)	Average Demand (60%)	High Demand (10%)	Expected Cost
62	$3,228,600	$3,414,400	$3,814,400	$3,398,660
67	$3,347,400	$3,272,000	$3,670,400	$3,334,460
73	$3,570,600	$3,504,000	$3,504,000	$3,523,980

The downside is that Kenwood Clocks must pay these workers salaries even if they do not work the full 8-h regular time shift under low- and average-demand scenarios. The expected cost of having 73 workers is $3,523,980, which is higher than that of having 62 workers. The expected cost of having 67 workers is $3,334,460, which is the lowest among the three. Therefore, Kenwood Clocks would be better off hiring another seven workers, assuming that the business remains stable in the long run.

Case Study 3.5.4

TTR Technologies is a manufacturer of thermal transfer ribbons. One of the raw materials for its product is a specialty chemical produced by Kumba Chemicals. There are no suitable substitutes for this chemical in the open market. Therefore, TTR has developed a partnership agreement with Kumba Chemicals, where the two companies share certain production-related information. TTR currently has 120 workers. Each month these workers work an average of 160 regular time hours and are willing to work up to 40 overtime hours. The salary for each worker is $3000 per month. The overtime hourly rate is $25 per worker. Producing one case of ribbons requires 10 worker hours. The inventory holding cost and backlog cost are $50 and $100 per case per month, respectively. TTR has made an aggregate plan for the July to December period as shown in Table 3.10. This plan is shared with Kumba Chemicals so that Kumba can better plan its supply of the specialty chemical. At the end of July, there is a major accident at Kumba Chemicals' plant. Kumba Chemicals estimated that its production will be adversely affected in August and September and can only supply enough specialty chemical for the production of 2800 cases of ribbons. After September things will go back to normal and it can supply whatever quantity of specialty chemical that is needed by TTR. How can you help TTR deal with this unexpected situation?

TABLE 3.10

TTR's Aggregate Plan for the July to December Period

Month	Demand	Production	Overtime	Inventory	Backlog
				0	0
July	1200	1760	0	560	0
August	2000	1920	0	480	0
September	2400	1920	0	0	0
October	1500	1780	0	280	0
November	2200	1920	0	0	0
December	1800	1800	0	0	0

TTR's current aggregate plan calls for the production of 1920 cases of ribbons in both August and September, which is a total of 3840 for the 2-month period. Because of the accident in Kumba Chemicals' plant, TTR will only have enough specialty chemical to produce a total of 2800 cases of ribbons in August and September. Therefore, we need to update our linear programming model to develop a new aggregate plan. Note that at the end of July TTR has build up an inventory of 560 cases of ribbons. Therefore, in our model, we need to plan for the 5-month period (from August to December) by adding a constraint on total production in August and September and setting the inventory at the beginning of August to 560. The linear programming model file for Gurobi Optimizer is as follows:

```
Minimize

\minimize the sum of labor cost, inventory holding cost, and backlog cost

   15000 W \regular time labor cost
   + 250 O2 + 250 O3 + 250 O4 + 250 O5 + 250 O6 \overtime labor cost
   + 50 I2 + 50 I3 + 50 I4 + 50 I5 + 50 I6 \inventory holding cost
   + 100 B2 + 100 B3 + 100 B4 + 100 B5 + 100 B6 \backlog cost

Subject To

\constraint due to shortage of raw material in August and September
P2 + O2 + P3 + O3 <= 2800

\regular time capacity constraint
16 W - P2 >= 0
16 W - P3 >= 0
16 W - P4 >= 0
16 W - P5 >= 0
16 W - P6 >= 0

\overtime capacity constraint
4 W - O2 >= 0
4 W - O3 >= 0
4 W - O4 >= 0
4 W - O5 >= 0
4 W - O6 >= 0

\inventory balance
I1 + P2 + O2 - I2 - B1 + B2 = 2000
I2 + P3 + O3 - I3 - B2 + B3 = 2400
I3 + P4 + O4 - I4 - B3 + B4 = 1500
I4 + P5 + O5 - I5 - B4 + B5 = 2200
I5 + P6 + O6 - I6 - B5 + B6 = 1800

\initial inventory and backlog
I1 = 560
B1 = 0

\ending inventory and backlog
I6 = 0
B6 = 0

\workforce size
W = 120

Bounds

\default is >= 0

General

\variables with known values do not need to be specified
   W P2 P3 P4 P5 P6 O2 O3 O4 O5 O6 I2 I3 I4 I5 B1 B2 B3 B4 B5

End
```

TABLE 3.11

Updated Aggregate Plan for TTR

Month	Demand	Production	Overtime	Inventory	Backlog
July				560	0
August	2000	1440	0	0	0
September	2400	1360	0	0	1040
October	1500	2400	480	0	140
November	2200	2220	300	0	120
December	1800	1920	0	0	0

The updated aggregate plan is shown in Table 3.11. We can see that TTR needs to backlog demand during September, October, and November. Workers are required to work overtime during October and November. Obviously TTR will incur higher cost due to the disruption in raw material supply. The lessons learned from this case study are

- We should work with upstream suppliers and downstream customers to ensure that inputs to our aggregate planning model are as accurate as possible.
- We need to update an aggregate plan when the assumptions made are no longer valid or new data emerges (e.g., change in demand).

Problem: Plastic Bottle Inc.

Plastic Bottle Inc. (PBI) makes PET (polyethylene terephthalate) bottles. PET bottles are produced using a "two-stage reheat and blow" process. In the first stage, amorphous PET is injection molded into a carefully designed preform, which is compact, rugged, and can be transported economically. In the second stage, the amorphous preform is passed through a multi-stage infrared oven where it is reheated to approximately 105°C and blown into a bottle.

PBI currently has 12 injection molding and 32 blow molding machines. On average, each machine incurs $24,000 of maintenance and depreciation cost per year. Plant maintenance requires shutting down all the machines. It is scheduled once a year and is performed by Molding Technology LLC. Due to staffing constraints, Molding Technology LLC can only perform maintenance in March, which requires shutting down the plant for 2.5 working days, or in July, which requires shutting down the plant for 5 working days.

Each injection molding machine requires one operator and can produce 0.5 thousand pounds of PET preform per hour. The material cost for 1000 lb of PET preform is $30. Each blow molding machine requires one operator

TABLE 3.12

PET Bottle Demand Forecast (in 1000 lb)

Month	Demand	Month	Demand
January	800	July	1100
February	1000	August	600
March	1550	September	850
April	1250	October	700
May	800	November	3000
June	900	December	1000

and can produce 0.2 thousand pounds of PET bottle per hour. Currently, PBI has 36 skilled operators that can operate both types of machines. Each operator is paid a monthly salary of $3200 and works a maximum of 8 h/day. There are 20 working days each month. In addition, each operator can work a maximum of 40 h of overtime a month, with an hourly pay of $30. A new operator requires 80 h of training before he/she can start operating a machine. Laying off an operator costs $4500.

The inventory holding cost is $10 per month for 1000 lb of PET preform and $50 per month for 1000 lb of PET bottle. The backlog cost of PET bottle is $70 per thousand pounds per month. PBI strives to provide the best-quality products to its customer. It evaluates all potential subcontractors and found that a number of subcontractors can provide high-quality PET preform at a cost of $80 (including raw material cost) per thousand pounds. However, no subcontractors can provide PET bottles at a level of quality that is acceptable to PBI.

The forecasted demand for the next year is shown in Table 3.12. PBI has no inventory and backlog at the beginning of the year. It does not want to hold any inventory or maintain any backlog at the end of the year.

Develop an aggregate plan for PBI and explain how you make the following decisions:

- When to build up PET preform inventory and when to purchase it from subcontractors?
- When to schedule plant maintenance?
- What is the cost that PBI is expected to incur?

Exercises

3.1   Bell Inc. needs to develop a production plan for laptop computers. The opening inventory is 100 laptops and the company wants to reduce the inventory to 80 by the end of the planning period. The company wants to keep the production volume constant over the planning period. Backlogs are not allowed. The expected demand for laptops is as follows:

Period	1	2	3	4	5	Total
Demand	110	120	130	120	120	600

How much should be produced in each period? What is the ending inventory in each period?

3.2 T & K Distribution is a distributor of herbal medicines. It imports herbal medicines in bulk and repackage them for sale to retail stores. Each worker can package 1000 bottles of herbal medicines a day. The inventory holding cost and backlog cost are $1 and $2 per month per bottle, respectively. The company has 20 employees; each of them earns $2000 a month. The demand (in thousands of bottles) and number of working days for the January to June period are shown as follows:

Month	January	February	March	April	May	June
Demand	400	400	620	520	420	400
Working days	24	23	27	24	26	26

With current workforce, develop an aggregate plan when no backlogs are allowed. Will T & K Distribution save money by allowing backlogs?

3.3 Refer to the data in Exercise 3.2, T & K Distribution is considering the use of temporary workers during high-demand period. Temporary workers can be hired from a local staffing company. The cost for hiring a temporary worker is $2800 per month plus a one-time fee of $1000. If temporary workers are as efficient as regular employees, can T & K reduce its cost by hiring temporary workers?

3.4 TDTech is a manufacturer of portable 3D scanners. Each unit of the scanner is equipped with a tripod and requires 10 worker hours to produce. The company has 50 hourly employees. Each of them can work a maximum of 160 h/month. The average hourly pay is $22 per employee. The material cost for producing one unit of scanner (including the tripod) is $500. The company is now considering purchasing the tripod from a supplier at a cost of $70 each. This would reduce the material cost for producing one unit of scanner to $480 and cut the production time to eight worker hours. Will TDTech be better off by purchasing tripods from the supplier instead of making it in-house?

3.5   Refer to Exercise 3.4, the demand for TDTech in the next 6 months is shown as follows:

Month	1	2	3	4	5	6
Demand	700	600	800	1000	900	800

The inventory holding cost for a unit of scanner is $10 per month. No backlogs are allowed. No employees will be hired or laid off during the planning horizon. Should TDTech consider purchasing tripods from the supplier? If yes, when should TDTech purchase tripods and in what quantity?

References

Fylstra, D., L. Lasdon, J. Watson, and A. Waren. 1998. Design and use of the Microsoft Excel solver. *Interface* **28**: 29–55.

Gurobi Optimization. 2012. Gurobi optimizer example tour version 5.0. http://www. gurobi.com/documentation/5.0/example-tour/node138 (accessed August 12, 2012).

Hanssmann, F. and S. W. Hess. 1960. A linear programming approach to production and employment scheduling. *Management Technology* **1**: 46–51.

Sipper, D. and R. L. Bulfin Jr. 1997. *Production: Planning, Control and Integration.* New York: McGraw-Hill.

4

Satisfying Customer Demand: Inventory Management

4.1 Overview

Inventory is a quantity of product units held by a company for some time to satisfy customer demand. It is a buffer between the supply and demand processes. The supply process contributes product units to the inventory, whereas the demand process depletes the same inventory. Inventory is necessary because of differences in rates and timing between supply and demand. Inventory management aims to determine the appropriate level of inventory held by a company. It is an important factor that impacts supply chain profitability and competitiveness. A number of factors influence the appropriate level of inventory. These factors are summarized as follows:

- *Economies of scale.* There are certain costs associated with purchasing and production of products. To reduce the average unit cost, a large quantity of product units may be purchased or produced. Quantity discount in product pricing also makes it desirable to purchase a large quantity of product units in each lot. These large lot sizes are ordered infrequently and placed in inventory to satisfy future demand.

- *Demand uncertainty.* Actual customer demand may exceed forecasted demand. To avoid the prospect of running out of stock, extra inventory, called *safety inventory*, may be held.

- *Customer satisfaction.* Inventory is built up so that customer demand can be met immediately. The more inventory is held, the higher the chance that customer demand can be met, which results in a higher level of customer satisfaction. However, more inventory held means higher inventory holding cost.

- *Price fluctuation.* Price fluctuation in the market may justify holding more inventory. However, this is highly speculative and should be left to the financial division of a company to handle.

There are several types of costs associated with inventory. *Purchasing cost*, denoted as c, is the per-unit cost paid to the supplier of the product. When a supplier offers quantity discount, the purchasing cost is a function of the order quantity, denoted as Q. *Ordering cost*, denoted as O, is the cost associated with placing an order with a supplier. For modeling purpose, this cost is assumed to be independent of the lot size (i.e., order quantity) and is often called *fixed ordering cost*. Its original formulation is attributed to Harris (1913) in a manufacturing situation where machines need to be set up for the production run. No matter how large the lot size is, the setup cost remains the same. In a purchasing situation, the fixed ordering cost may be the cost related to placing and receiving the order.

Inventory ties up capital, requires storage and handling, may need insurance, may incur tax, and may be damaged or become obsolete. All of these cost money. This is called *inventory holding cost*, denoted as H. When a customer demand cannot be satisfied, *understocking cost*, denoted as C_u, is incurred. The understocking cost could be the lost profit and goodwill if the demand is lost. If the demand is backlogged, the understocking cost could be the discount offered to keep the customer and/or costs associated with record keeping and expedited shipment to the customer. The goal of inventory management is to minimize the costs associated with inventory while satisfying customer demand.

4.2 Economic Order Quantity

4.2.1 Basic Concept of Economic Order Quantity

Strickland Propane purchases tanks of propane from AmeriGas and delivers them to customer's homes. The demand is constant at 50 tanks per week. The purchasing cost for a tank of propane is $12. The ordering cost is $500 per order. The inventory holding cost is $3 per tank per year. How often should propane be ordered from AmeriGas and in what quantity?

To answer this question, we need to understand the relationship between ordering frequency/order quantity and various costs incurred by Strickland Propane. There are three categories of cost, namely, purchasing cost, ordering cost, and inventory holding cost. We will analyze these three categories of costs on a yearly basis, assuming there are 52 weeks in a year. The total yearly demand, denoted as D, is 2600 (50×52) tanks of propane. Let f denote the ordering frequency (number of orders per year) and Q denote the order quantity. The ordering policy is such that the total yearly demand is always satisfied, that is, $f \times Q = D$. Figure 4.1 shows the inventory profile under this circumstance, assuming a linear inventory depletion rate. One can see that it takes $1/f$ (or Q/D) years for the entire lot of propane to be sold. When the

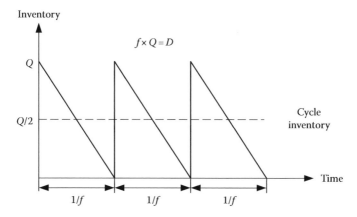

FIGURE 4.1
Inventory profile.

inventory level reaches zero, Strickland Propane receives another lot of Q tanks of propane. This process of depleting and receiving inventory repeats itself. The average inventory, called *cycle inventory*, is thus $Q/2$.

Strickland Propane needs to order 2600 tanks of propane every year in order to satisfy customer demand. Therefore, the yearly purchasing cost is $2,600 \times 12 = \$31,200$, which is independent of the ordering frequency/order quantity. On the other hand, the yearly inventory holding cost is $Q/2 \times 3$, which is a function of Q. To reduce the inventory holding cost, a smaller Q is preferred. However, the yearly ordering cost is $f \times 500 = (D/Q) \times 500 = (2600/Q) \times 500$, which increases when Q is reduced. Therefore, we need to consider the sum of ordering cost and inventory holding cost when determining Q. Figure 4.2 plots the yearly ordering cost, yearly inventory holding cost, and the sum of the two (total cost per year) as a function of the order quantity Q. As Q increases, the yearly ordering cost decreases, whereas the yearly inventory holding cost increases. It is clear that we need to make a trade-off between

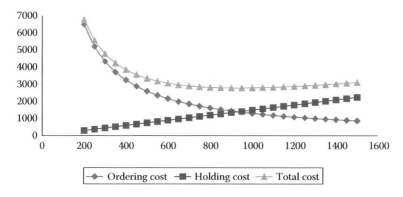

FIGURE 4.2
Cost as a function of order lot size.

the yearly ordering cost and the yearly inventory holding cost in order to find the order quantity that minimizes the total yearly cost.

Recall that we use O to denote the fixed ordering cost (per order) and H to denote the inventory holding cost (per unit). Therefore, the total yearly cost that we seek to minimize is

$$TC = \left(\frac{D}{Q}\right)O + \left(\frac{Q}{2}\right)H \qquad (4.1)$$

To find the value of Q that minimizes TC, we take the derivative of TC against Q and set it to zero, as follows:

$$\frac{d(TC)}{dQ} = \frac{d\left(\dfrac{DO}{Q}\right)}{dQ} + \frac{d\left(\dfrac{QH}{2}\right)}{dQ} = -\frac{DO}{Q^2} + \frac{H}{2} = 0 \qquad (4.2)$$

Solving Equation 4.2 we have the optimal order quantity, denoted as Q^*, as follows:

$$Q^* = \sqrt{\frac{2DO}{H}} \qquad (4.3)$$

This optimal order quantity is called the *Economic Order Quantity (EOQ)*. Therefore, Strickland Propane should order $Q^* = \sqrt{(2 \times 2600 \times 500)/3} \approx 931$ tanks of propane every time it places an order. The optimal ordering frequency is $f^* = D/Q^* \approx 2.8$ orders every year, or once every 18.6 weeks. The total yearly inventory cost (ordering cost and inventory holding cost) is $(2600/931) \times 500 + (931/2) \times 3 \approx \$2793$.

4.2.2 *EOQ* Calculation under Different Scenarios

Using the *EOQ* equation is quite simple when the inventory holding cost is given. In the real world, how to determine the inventory holding cost may not be straightforward. In general, the inventory holding cost has two major components, namely, capital cost and storage cost. Capital cost, denoted as I, is usually calculated as a percentage of the purchasing cost c, that is, $I = \alpha c$. For example, instead of buying a tank of propane and holding it for a year, Strickland Propane can invest the \$12 and receive 5% interest ($\alpha = 5\%$). Therefore, the capital cost for holding a tank of propane for a year is \$0.6 (12 × 5%). Suppose full propane tanks are stored in a public storage facility that charges \$0.2 per tank per month, whereas empty tanks are stored openly outside Strickland Propane and do not incur storage cost.

Then the storage cost, denoted as W, for holding a full tank of propane for a year is $2.4 ($0.2 \times 12$). Therefore, the inventory holding cost for a tank of propane is $3 ($H = I + W = 0.6 + 2.4$) per year. Now we look at three different scenarios when calculating *EOQ*.

Scenario 1: The public storage company is offering an option of leasing a fixed storage area to Strickland Propane to store the propane tanks. The cost is $18 per sq. ft. per year. The lease is on a yearly basis. For example, if Strickland Propane decides to lease 100 sq. ft. of storage area, then it incurs a cost of $1800 a year even if it does not always fully utilize the storage area. It is estimated that 1 sq. ft. of storage area can hold 10 propane tanks. Can Strickland Propane save money by leasing a fixed storage area? How big a storage area is needed?

Because 1 sq. ft. of storage space can hold 10 propane tanks and costs $18 per year, the storage cost for a tank of propane is $1.8 ($18 \div 10$) per year. This is lower than the $2.4 per year Strickland Propane currently incurs. However, currently the yearly storage cost depends on the average number of tanks of propane in storage; yet the new option requires a fixed yearly storage cost. We will need to reevaluate the total yearly inventory cost by separating capital cost from storage cost. Our capital cost is $(Q/2)I$ because our cycle inventory is still $Q/2$. However, we need to lease a storage area that can hold Q tanks of propane because our order quantity is Q. Therefore, our yearly storage cost is QW, where $W = 1.8$. The total cost to be minimized is thus

$$TC = \left(\frac{D}{Q}\right)O + \left(\frac{Q}{2}\right)I + QW \qquad (4.4)$$

Again, we take the derivative of TC against Q and set it to zero, as follows:

$$\frac{d(TC)}{dQ} = \frac{d\left(\dfrac{DO}{Q}\right)}{dQ} + \frac{d\left(\dfrac{QI}{2}\right)}{dQ} + \frac{d(QW)}{dQ} = -\frac{DO}{Q^2} + \frac{I}{2} + W = 0 \qquad (4.5)$$

Solving Equation 4.5 we have the optimal order quantity

$$Q^* = \sqrt{\frac{2DO}{I + 2W}} \qquad (4.6)$$

Therefore, if Strickland Propane chooses to lease a fixed storage area, the area should be large enough to hold $Q^* = \sqrt{(2 \times 2600 \times 500)/(0.6 + 2 \times 1.8)} \approx 787$ tanks of propane. The total yearly inventory cost is $(2600/787) \times 500 + (787/2) \times 0.6 + 787 \times 1.8 \approx \$3305$. This is more than the current total yearly inventory cost. Therefore, it does not make sense to lease a fixed storage area.

In this scenario, the storage space is dedicated to store Strickland Propane's propane tanks. Refer to the inventory profile shown in Figure 4.1; we can see that during each cycle the storage space will initially hold Q tanks of propane and ends up holding 0 tanks of propane. On average, only half of the storage area is utilized. The storage cost of $1.8 per tank per year we previously calculated assumes that the storage space is fully utilized. Because in reality the utilization of the storage space is 50%, the actual storage cost for a tank of propane is $1.8 \times 2 = \$3.6$ per year. If we let $H = I + 2W$, then Equation 4.6 reduces to the basic EOQ equation. We can quickly find out that by leasing a fixed storage area the inventory holding cost is $H = 0.6 + 2 \times 1.8 = \$4.2$. It is higher than the current inventory holding cost of $3. That is why it does not make sense to lease a fixed storage area.

Scenario 2: Strickland Propane recently built a storage area outside its store that can hold 600 tanks of propane. Additional full tanks of propane are stored in the public storage facility. Empty tanks are still stored openly at Strickland Propane. As such, marginal storage cost is incurred only when the quantity of propane on hand exceeds 600 tanks. Can Strickland Propane change the ordering policy to save money?

This case is a little tricky because the inventory holding cost is no longer a linear function of the order quantity. Let A denote the threshold of quantity above which a storage cost is incurred. The fraction of time in a year that a storage cost is incurred is $(Q-A)/Q$ and the average inventory during this time period is $(Q-A)/2$, as shown in Figure 4.3. Therefore, the total cost to be minimized is

$$TC = \left(\frac{D}{Q}\right)O + \left(\frac{Q}{2}\right)I + \frac{(Q-A)}{Q} \times \frac{(Q-A)}{2}W \qquad (4.7)$$

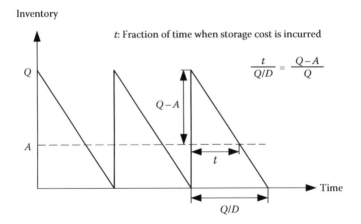

FIGURE 4.3
Analysis of threshold-based storage cost.

The *EOQ* in this case is

$$Q^* = \sqrt{\frac{2DO + WA^2}{I + W}} = \sqrt{\frac{2DO + WA^2}{H}} \qquad (4.8)$$

Therefore, Strickland Propane should order $Q^* = \sqrt{(2DO + WA^2)/(I + W)} = \sqrt{(2 \times 2600 \times 500 + 2.4 \times 600^2)/(0.6 + 2.4)} \approx 1075$ tanks of propane each time. The total yearly inventory cost is $(2600/1075) \times 500 + (1075/2) \times 0.6 + (1075 - 600)/1075 \times (1075 - 600)/2 \times 2.4 \approx \$1637$.

Scenario 3: A local transportation company offers to handle the ordering and shipping of propane for Strickland Propane. The company proposes to charge a fixed fee of $200 per shipment if the order quantity is no more than 800 tanks. For any tanks above 800, the company charges an additional $2 per tank. Should Strickland Propane accept this offer?

If Strickland Propane orders 1075 tanks of propane each time, it will not make sense to use the transportation company because it will need to pay $750 (200 + 275 × 2) per order. However, the fixed ordering cost has changed from $500 to $200 when the lot size is less than 800. Therefore, Strickland Propane should reevaluate the *EOQ*, which is

$$Q^* = \sqrt{(2DO + WA^2)/(I + W)} = \sqrt{(2 \times 2600 \times 200 + 2.4 \times 600^2)/(0.6 + 2.4)} \approx 797.$$

Because $Q^* < 800$, Strickland Propane can take advantage of the lower fixed ordering cost. The total yearly inventory cost is $(2600/797) \times 200 + (797/2) \times 0.6 + (797 - 600)/797 \times (797 - 600)/2 \times 2.4 \approx \$950$, a saving of $687 compared to the option of in-house ordering and shipping. Therefore, Strickland Propane should use the local transportation company.

4.2.3 Multiple Types of Products

A company usually needs to handle multiple types of products. A portion of the ordering cost is independent of the product variety (e.g., transportation cost for a truckload of products). The rest is related to each individual type of products (e.g., loading from each product supplier). We now investigate the ordering strategy under this circumstance. The following notations are used:

n: the types of products

i: index of product type, $i = 1, 2, 3,..., n$

D_i: annual demand for product i

O: ordering cost incurred each time an order is placed (independent of product variety)

o_i: additional ordering cost incurred if product i is included in the order

H_i: annual inventory holding cost for product i

We have three ordering strategies: (1) order each type of products separately, (2) order all the products jointly, and (3) order a subset of the products jointly. These strategies are called *no aggregation, complete aggregation,* and *tailored aggregation,* respectively (Chopra and Meindl 2003). The no aggregation strategy is likely to result in high cost. The complete aggregation strategy will likely produce some cost savings. However, it aggregates low-demand and high-demand products. When the product-specific ordering cost (o_i) for a low-demand product is high, this strategy may still result in high total cost. Under this circumstance, it may be better to use the tailored aggregation strategy. Therefore, it is necessary to evaluate all three strategies and compare the total inventory cost before making a decision.

When using the no aggregation strategy, we simply use the basic *EOQ* equation, that is, Equation 4.3, and use $O + o_i$ as the fixed ordering cost for product i. We then calculate the ordering frequency/order quantity for each type of product separately. When using the complete aggregation strategy, let f be the number of orders placed per year. Our total annual cost is

$$TC = \left(O + \sum_{i=1}^{n} o_i \right) f + \frac{\sum_{i=1}^{n} D_i H_i}{2f} \tag{4.9}$$

By taking the derivative of *TC* against f and set it to zero, we can obtain the optimal ordering frequency as

$$f^* = \sqrt{\frac{\sum_{i=1}^{n} D_i H_i}{2\left(O + \sum_{i=1}^{n} o_i \right)}} \tag{4.10}$$

Therefore, our ordering frequency is f^* times per year and the order quantity for product i is D_i/f^*.

The tailored aggregation strategy is more complicated and requires the following procedure:

Step 1: Find the most frequently ordered product, assuming each product is ordered independently. Calculate $f_i = \sqrt{D_i H_i/2(O + o_i)}$, $i = 1, 2, 3, \ldots, n$; let $f_m = \max\{f_i\}$. Product m, is the most frequently ordered product.

Step 2: Identify the ordering frequency of other products as a relative multiple of the most frequently ordered product. It is assumed that product i is ordered jointly with product m so only the ordering cost specific to product i is considered when calculating the ordering frequency, that is, $f_i' = \sqrt{D_i H_i/2o_i}$, $i = 1, 2, 3, \ldots, n$; and $i \neq m$. Let $r_i = \lceil f_m/f_i' \rceil$, $i \neq m$ and $r_m = 1$.

Step 3: Recompute the frequency of the most frequently ordered product, taking into account joint ordering. Note that product i is ordered in every r_i order. Therefore, the annual ordering cost is $f_m\left(O + \sum_{i=1}^{n} o_i / r_i\right)$ and the annual inventory holding cost is $\sum_{i=1}^{n} H_i D_i r_i / 2 f_m$. Thus, the optimal ordering frequency that minimizes the total cost is $f_m^* = \sqrt{\sum_{i=1}^{n} D_i H_i r_i \Big/ 2\left(O + \sum_{i=1}^{n} o_i / r_i\right)}$.

Step 4: Recompute the ordering frequency of other products $f_i^* = f_m^* / r_i, i \neq m$.

We now look at an example of inventory management for multiple products. Pop Electronics sells laptop computers, desktop computers, and HDTVs, and the annual demands are 52,000, 26,000, and 5,200, respectively. The annual inventory holding cost for a unit of laptop computer, a unit of desktop computer, and a unit of HDTV are $48, $40, and $48, respectively. Product ordering and shipping are outsourced to Interstate Shipping. Interstate Shipping operates a distribution center. It transports products from individual suppliers to the distribution center and then transports products to its customers. It charges Pop Electronics $800, $1000, and $1400 each time it transports laptop computers, desktop computers, and HDTVs, respectively, from the suppliers to its distribution center. To transport products from the distribution center to Pop Electronics, Interstate Shipping charges $4000 irrespective of the variety and quantity of products. What ordering strategy should Pop Electronics use in order to minimize cost?

First, we use the no aggregation strategy. The results are summarized in Table 4.1. The total annual cost for the three products is $308,695. We then use the complete aggregation strategy and compute $f^* = \sqrt{\sum_{i=1}^{n} D_i H_i \Big/ 2\left(O + \sum_{i=1}^{n} o_i\right)} =$

$\sqrt{(52,000 \times 48 + 26,000 \times 40 + 5,200 \times 48)/(2 \times (4,000 + 800 + 1,000 + 1,400))} = 16.21$.

The total annual cost is $TC = \left(O + \sum_{i=1}^{n} o_i\right) f^* + \sum_{i=1}^{n} D_i H_i / 2 f^* = (4,000 + 800 +$

$1,000 + 1,400) \times 16.21 + (52,000 \times 48 + 26,000 \times 40 + 5,200 \times 48) / (2 \times 16.21) =$ $233,479. This total annual cost is lower than that with no aggregation.

TABLE 4.1

No Aggregation Strategy for Inventory Management at Pop Electronics

	Laptop	Desktop	HDTV
Demand	52,000	26,000	5,200
Holding cost	$48	$40	$48
Ordering cost	$4,000 + $800	$4,000 + $1,000	$4,000 + 1,400
EOQ	3,225	2,550	1,082
Annual cost	$154,795	$101,980	$51,920

Now we use the tailored aggregation strategy as follows:

Step 1: We have $f_{laptop} = \sqrt{(52,000 \times 48)/(2(4,000+800))} = 16.12$,

$$f_{desktop} = \sqrt{(26,000 \times 40)/(2(4,000+1,000))} = 10.20,$$

$$f_{HDTV} = \sqrt{(5200 \times 48)/(2(4000+1400))} = 4.81.$$

Laptop computer is the most frequently ordered product.

Step 2: $f'_{desktop} = \sqrt{(26,000 \times 40)/(2 \times 1,000)} = 22.80$,

$$f'_{HDTV} = \sqrt{(5200 \times 48)/(2 \times 1400)} = 9.44, \ r_{desktop} = \lceil 16.12/22.80 \rceil = 1,$$

$$r_{HDTV} = \lceil 16.12/9.44 \rceil = 2, \ r_{laptop} = 1.$$

Step 3: $f^{*}_{laptop} = \sqrt{(48 \times 52,000 \times 1 + 40 \times 26,000 \times 1 + 48 \times 5,200 \times 2)/}$

$$\sqrt{(2(4,000+(800/1)+(1,000/1)+1,400/2))} = 17.62.$$

Step 4: $f^{*}_{desktop} = 17.62/1 = 17.62, f^{*}_{HDTV} = 17.62/2 = 8.81.$

The cost calculation for the tailored aggregation strategy is shown in Table 4.2, with the common ordering cost (transportation from distribution center to Pop Electronics) allocated to laptop computers. The total annual cost is \$229,037, which is even lower than that with complete aggregation. Therefore, we should use this tailored aggregation strategy to manage the inventory at Pop Electronics.

4.2.4 Quantity Discounts

In business-to-business transactions, suppliers may entice their customers to place larger orders by offering quantity discounts. If the quantity purchased

TABLE 4.2

Tailored Aggregation Strategy for Inventory Management at Pop Electronics

	Laptop	Desktop	HDTV
Order frequency	17.62	17.62	8.81
Order quantity	2,591	1,476	590
Ordering cost	\$84,576	\$17,620	\$12,334
Holding cost	\$70,829	\$29,512	\$14,166
Annual cost	\$155,405	\$47,132	\$26,500

is greater than a specified "price break" quantity, the unit price is reduced. Naturally, a customer would like to take advantage of this offer to purchase a larger quantity of products, especially when the product is used continuously. However, a larger quantity means a larger inventory and thus higher inventory holding cost. Therefore, we need to analyze whether savings gained by purchasing a larger quantity of products can offset the additional inventory holding cost.

There are two types of quantity discounts. *All-unit discount* applies the discounted price to all units, whereas *incremental discount* applies the discounted price only to those units over the price break quantity. We use the following notations to describe these situations:

m: the number of price breaks

i: index of price break intervals, $i = 1, 2, ..., m$

q_i: the upper limit of the ith price break interval

c_i: the unit cost in the ith price break interval $[q_{i-1}, q_i]$

In general, we have $q_0 = 0$, $q_m = \infty$, and $c_{i+1} < c_i$. For all-unit discount, the total price for purchasing Q_i (where $q_{i-1} \leq Q_i \leq q_i$) units is

$$P(Q_i) = c_i Q_i \tag{4.11}$$

On the other hand, for incremental discount, the total price is

$$P(Q_i) = \sum_{j=1}^{i-1} c_j(q_j - q_{j-1}) + c_i(Q_i - q_{i-1}) \tag{4.12}$$

Here, we use an example to illustrate these two types of quantity discount. Bell Inc. manufactures laptop computers. The company purchases batteries from its suppliers. The annual demand is 9600 batteries. The fixed ordering cost is $500. The annual capital cost is 10% of the unit cost, and the annual storage cost is $12 per unit. All components are stored in a common warehouse. Longey and Everlast are two battery manufacturers. Their product and service are equal, so batteries will be purchased solely based on cost. Both suppliers offer quantity discount. For Longey, if the quantity is less than 500 units, the price is $100 per unit; if the quantity is 500 units or more but less than 1000 units, the price is $90 per unit; if the quantity is 1000 units or above, the price is $80 per unit. For Everlast, if the quantity is less than 500 units, the price is $90 per unit; if the quantity is 500 units or more but less than 1000 units, the price for those units above 500 is $80 per unit; if the quantity is 1000 units or above, the price for those units above 1000 is $65 per unit. For both companies, the price break intervals are the same. Specifically, there are three price breaks ($m = 3$) with $q_1 = 500$ and $q_2 = 1000$. However, the two

TABLE 4.3

Battery Price Schedules from Everlast and Longey

Quantity	Longey	Everlast
$0 \leq Q < 500$	$100Q$	$90Q$
$500 \leq Q < 1000$	$90Q$	$90 \times 500 + 80 \times (Q - 500)$
$1000 \leq Q < \infty$	$80Q$	$90 \times 500 + 80 \times 500 + 65 \times (Q - 1000)$

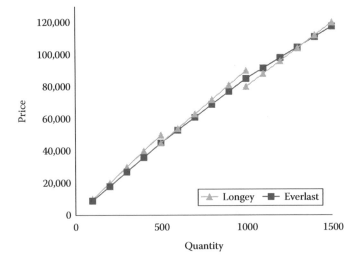

FIGURE 4.4
Comparison of price schedules from Everlast and Longey.

price schedules are different. They are summarized as shown in Table 4.3 and graphically illustrated in Figure 4.4.

As previously mentioned, the inventory holding cost consists of two components, namely, capital cost and storage cost. The capital cost is a function of the unit cost. Therefore, *EOQ* is also a function of the unit cost. Because the unit cost changes when order quantity changes, material cost must be included when trying to find the *EOQ* to minimize the total cost.

Let us look at all-unit discount first. Let Q_i^* denote the *EOQ* calculated given a unit cost of c_i, the total annual cost (including material cost, ordering cost, and holding cost) is

$$TC(Q_i^*) = Dc_i + \left(\frac{D}{Q_i^*} \right) O + \frac{Q_i^*}{2} H \qquad (4.13)$$

Note that Equation 4.13 is valid only when $q_{i-1} \leq Q_i^* < q_i$. If $Q_i^* < q_{i-1}$, we should try to raise the order quantity to q_{i-1} to take advantage of the lower unit cost. However, we need to confirm that by doing so we can lower the total annual cost. We can start with the lowest unit cost to perform this analysis. The procedure is as follows:

Step 1: Let $Q^* = 0$, $TC^* = \infty$, $i = m$.

Step 2: Calculate Q_i^*; if $Q_i^* < q_{i-1}$, let $Q_i^* = q_{i-1}$, $TC(Q_i^*) = Dc_i + (D/Q_i^*)O + (Q_i^*/2)H$. Otherwise go to Step 4.

Step 3: If $TC(Q_i^*) < TC^*$, let $Q^* = Q_i^*$ and $TC^* = TC(Q_i^*)$. Let $i = i - 1$; go to Step 2.

Step 4: Let $TC(Q_i^*) = Dc_i + (D/Q_i^*)O + (Q_i^*/2)H$ if $TC(Q_i^*) < TC^*$, let $Q^* = Q_i^*$ and $TC^* = TC(Q_i^*)$.

Step 5: Stop; the optimal order quantity is Q^* with total cost TC^*.

Now we follow this procedure to determine the optimal order quantity and total annual cost if Longey is chosen as the supplier.

Step 1: Let $Q^* = 0$, $TC^* = \infty$, $i = 3$

Step 2: Calculate EOQ with $c_3 = \$80$.

$$Q_3^* = \sqrt{(2DO/(I+W))} = \sqrt{((2 \times 9600 \times 500)/(80 \times 10\% + 12))} \approx 693;$$

$$\text{because } Q_3^* < q_2, \text{ let } Q_3^* = q_2 = 1000,$$

$$TC(Q_3^*) = 9600 \times 80 + (9600/1000) \times 500 + (1000/2) \times (80 \times 10\% + 12) \approx \$782,800.$$

Step 3: Because $TC(Q_3^*) < TC^*$, let $Q^* = 1,000$ and $TC^* = \$782,800$. Let $i = 2$; go to Step 2.

Step 2: Calculate the EOQ with $c_2 = \$90$.

$$Q_2^* = \sqrt{2DO/(I+W)} = \sqrt{(2 \times 9600 \times 500)/(90 \times 10\% + 12)} \approx 676;$$
$$\text{because } Q_2^* > q_1, \text{ go to Step 4.}$$

Step 4:

$$TC(Q_2^*) = 9,600 \times 90 + (9,600/676) \times 500 + (676/2) \times (90 \times 10\% + 12) \approx \$878,199.$$

Because $TC(Q_2^*) > TC^*$, do nothing.

Step 5: The optimal order quantity is 1,000 and the total cost is $782,800.

Now we look at incremental discount. In this case, the total annual cost for an order quantity of Q_i ($q_{i-1} \le Q_i < q_i$) is

$$TC(Q_i) = \frac{P(Q_i)}{Q_i}D + \left(\frac{D}{Q_i}\right)O + \left[W + \alpha \frac{P(Q_i)}{Q_i}\right]\left(\frac{Q_i}{2}\right) \qquad (4.14)$$

By taking the derivative of $TC(Q_i)$ against Q_i and set it to zero, we can find the EOQ as

$$Q_i^* = \sqrt{\frac{2D\left[O + \sum_{j=1}^{i-1} c_j(q_j - q_{j-1}) - c_i q_{i-1}\right]}{\alpha c_i + W}} \tag{4.15}$$

Note that when $i = 1$, Equation 4.15 reduces to $Q_i^* = \sqrt{2DO/(\alpha c_i + W)}$. Again, this optimal order quantity is valid only when $q_{i-1} \le Q_i^* < q_i$. It has been shown that even this condition holds, there is no guarantee that Q_i^* will result in overall minimum cost. In addition, the minimum cost point would never occur at a price break point (Hadley and Whitin 1963). Therefore, we have to investigate Q_i^* for each price break. The following procedure is used to find the optimal order quantity:

Step 1: Let $Q^* = 0$, $TC^* = \infty$, $i = 1$.

Step 2: Calculate Q_i^* based on Equation 4.15; if $q_{i-1} \le Q_i^* < q_i$, calculate $TC(Q_i^*)$ based on Equation 4.14. Otherwise let $TC(Q_i^*) = \infty$.

Step 3: Let $i = i + 1$. If $i \le m$, go to Step 2.

Step 4: Let $TC(Q_f^*) = \min\{TC(Q_i^*)\}$; then $Q^* = Q_f^*$ and $TC^* = TC(Q_f^*)$.

Now we follow this procedure to determine the optimal order quantity and total annual cost if Everlast is chosen as the supplier.

Step 1: Let $Q^* = 0$, $TC^* = \infty$, $i = 1$.

Step 2: $Q_1^* = \sqrt{2DO/(\alpha c_i + W)} = \sqrt{(2 \times 9600 \times 500)/(10\% \times 90 + 12)} \approx 676$; because $Q_1^* > q_1$, let $TC(Q_1^*) = \infty$.

Step 3: Let $i = 2$. Because $i \le 3$, go to Step 2.

Step 2:

$$Q_2^* = \sqrt{\frac{2D\left[O + \sum_{j=1}^{1} c_j(q_j - q_{j-1}) - c_2 q_1\right]}{\alpha c_2 + W}} = \sqrt{\frac{2 \times 9600 \times (500 + 90 \times 500 - 80 \times 500)}{10\% \times 80 + 12}}$$

≈ 2298; because $Q_2^* > q_2$, let $TC(Q_2^*) = \infty$.

Step 3: Let $i = 3$. Because $i \le 3$, go to Step 2.

$$\text{Step 2: } Q_3^* = \sqrt{\dfrac{2D\left[O + \displaystyle\sum_{j=1}^{2} c_j(q_j - q_{j-1}) - c_3 q_2\right]}{\alpha c_3 + W}}$$

$$= \sqrt{\dfrac{2 \times 9600 \times (500 + 90 \times 500 + 80 \times 500 - 65 \times 1000)}{10\% \times 65 + 12}} \approx 4613;$$

because $q_2 \leq Q_3^* < q_3$,

$$\text{let } TC(Q_i) = \dfrac{P(Q_i)}{Q_i}D + \left(\dfrac{D}{Q_i}\right)O$$

$$+ \left[W + \alpha\dfrac{P(Q_i)}{Q_i}\right]\left(\dfrac{Q_i}{2}\right) = \dfrac{500 \times 90 + 500 \times 80 + (4,613 - 1,000) \times 65}{4,613} \times 9,600$$

$$+ \left(12 + 10\% \times \dfrac{500 \times 90 + 500 \times 80 + (4,613 - 1,000) \times 65}{4,613}\right)$$

$$\times \dfrac{4,613}{2} \approx \$710,332.$$

Step 3: Let $i = 4$. Because $i > 3$, go to Step 4.

Step 4: $Q^* = Q_3^* = 4,613$ and $TC^* = \$710,332$.

This analysis showed that the total annual cost is lower when purchasing batteries from Everlast.

4.3 Safety Inventory

When demand is uncertain, a product shortage may occur if the actual demand exceeds the forecasted demand. Therefore, a company should maintain additional inventory, called *safety inventory*, to satisfy customer demand. More safety inventory decreases the chance that a customer faces a shortage but incurs more inventory holding cost. Therefore, we need to investigate how to determine an appropriate level of safety inventory.

4.3.1 Continuous Review

A number of factors influence the determination of an appropriate level of safety inventory. A major factor is inventory review policy. Note that managing inventory under demand uncertainty requires knowing the level of inventory on hand. Therefore, we must decide when to review inventory to determine its level. The ideal situation is that the level of inventory on hand is known at any

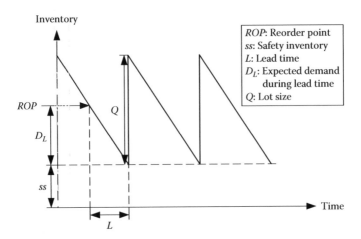

FIGURE 4.5
Inventory profile with safety inventory.

given point in time. This requires the implementation of a *continuous review* policy through the use of an automated inventory monitoring system. Under this circumstance, let us look at the inventory profile with safety inventory, as shown in Figure 4.5. Because the level of inventory is known at any given point in time, we can set a *reorder point* (ROP), which is the inventory level that triggers a replenishment order. The replenishment *lead time*, denoted as L, is the duration between when an order is placed and when it is received. The *safety inventory*, denoted as ss, is the difference between the reorder point and the expected demand during lead time, denoted as D_L. In other words, we have

$$ss = ROP - D_L \tag{4.16}$$

In order to model demand uncertainty, we assume that the demand during lead time follows a normal distribution with a mean D_L and a standard deviation σ_L. If the actual demand during lead time exceeds *ROP*, then we experience a stockout during the replenishment cycle, which means we are not able to meet all customer demands. The fraction of replenishment cycles that end with all the customer demands being met is called the *cycle service level* (CSL). In other words, the *CSL* is the probability that the actual demand during lead time is less or equal to *ROP*, that is,

$$CSL = P(x \leq ROP), \tag{4.17}$$

where x is the demand during lead time.

As illustrated in Figure 4.6, we have

$$CSL = F(ROP, D_L, \sigma_L) \quad \text{or} \quad ROP = F^{-1}(CSL, D_L, \sigma_L) \tag{4.18}$$

$$ss = F^{-1}(CSL, D_L, \sigma_L) - D_L = F_S^{-1}(CSL) \times \sigma_L \tag{4.19}$$

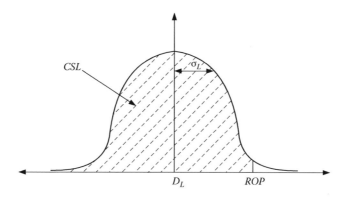

FIGURE 4.6
Relationship between cycle service level and reorder point.

where $F(x,\mu,\sigma)$ and $F^{-1}(x,\mu,\sigma)$ are the normal cumulative distribution function and inverse normal cumulative distribution function, respectively, with a mean μ and a standard deviation σ, and $F_S^{-1}(x)$ is the inverse standard normal cumulative distribution function.

Another measure of product availability is the product *fill rate* (*fr*), which is the fraction of product demand that is satisfied from inventory. To calculate the fill rate, we need to determine *expected shortage per replenishment cycle* (*ESC*) first. Let $f(x)$ be the probability density function of the demand distribution during lead time. Shortage occurs when $x > ROP$, that is, the demand during lead time is greater than the *ROP*. Therefore, the *ESC* is

$$ESC = \int_{x=ROP}^{\infty} (x - ROP)f(x)dx \tag{4.20}$$

From Equation 4.16, we have $ROP = D_L + ss$. Substituting $D_L + ss$ into Equation 4.20 for *ROP* and taking into account that $f(x)$ follows a norm distribution with a mean D_L and a standard deviation σ_L, we have

$$ESC = -ss\left[1 - F_S\left(\frac{ss}{\sigma_L}\right)\right] + \sigma_L f_S\left(\frac{ss}{\sigma_L}\right) \tag{4.21}$$

where
F_S is the standard normal cumulative distribution function
f_S is the standard normal density function

During each replenishment cycle, the expected product demand is the lot size *Q*. Product demand satisfied from inventory is $Q - ESC$. Therefore, the product fill rate is calculated as follows:

$$fr = \frac{Q - ESC}{Q} = 1 - \frac{ESC}{Q} \tag{4.22}$$

Note that both *CSL* and fill rate are functions of safety inventory. Specifically, both *CSL* and fill rate increase as the safety inventory increases. Fill rate is also a function of the order lot size. Given the same level of safety inventory, increasing the order lot size increases the fill rate. On the other hand, *CSL* is independent of the order lot size. In other words, CSL is an event-oriented performance measure, whereas fill rate is a quantity-oriented performance measure.

Now let us look at an example. Weekly demand of packs of AAA batteries in BVE store is normally distributed with a mean of 1000 and a standard deviation of 50. The replenishment lead time is 2 weeks. The order lot size is 3000. The desired *CSL* is 95%. What level of safety inventory should BVE carry? When should a replenishment order be placed? What is the fill rate given this inventory policy?

First, we need to calculate the average and standard deviation of the demand during lead time. Our weekly demand follows a normal distribution with $D_W = 1000$ and $\sigma_W = 50$. The lead time $L = 2$. Assuming weekly demands are independent, we have $D_L = D_W \times 2 = 1000 \times 2 = 2000$ and $\sigma_L = \sqrt{\sum_{i=1}^{L} \sigma_W^2} = \sqrt{L}\sigma_W = \sqrt{2} \times 50 = 70.7$. From Equation 4.18, we have $ROP = F^{-1}(CSL, D_L, \sigma_L) = F^{-1}(95\%, 2000, 70.7) \approx 2116$. Note that $F^{-1}(95\%, 2000, 70.7)$ can be quickly calculated in Microsoft Excel by typing "=NORMINV(0.95,2000,70.7)" in a cell and hitting the Enter key.

From Equation 4.16, we have $ss = ROP - D_L = 2116 - 2000 = 116$. Therefore, BVE should carry 116 packs of AAA batteries as safety inventory and place a replenishment order whenever the inventory level falls to 2,116. The expected shortage given this inventory policy is $ESC = -ss\left[1 - F_S(ss/\sigma_L)\right] + \sigma_L f_S(ss/\sigma_L) = -116 \times \left[1 - F_S(116/70.7)\right] + 70.7 \times f_S(116/70.7) = 1.49$. Therefore, the fill rate is $fr = 1 - (ESC/Q) = 1 - (1.49/3000) = 99.95\%$ Note that $F_S(116/70.7)$ and $f_S(116/70.7)$ can be quickly calculated in Microsoft Excel by typing "=NORMDIST(116/70.7,0,1,1)" and "=NORMDIST(116/70.7,0,1,0)" in a cell, respectively, and hitting the Enter key.

4.3.2 Periodic Review

The continuous review policy is feasible only when an automated inventory monitoring system is in place. Without such an inventory monitoring system, a company may choose to review its inventory level after a fixed period of time *T*. This is called a *periodic review* policy. After the review, an order is placed such that the level of current inventory plus the replenishment lot size equals a predetermined level called the *order-up-to level* (*OUL*). The inventory profile given a periodic review policy is shown in Figure 4.7. Assuming at time 0 the inventory on hand is at the *OUL*, so no replenishment order is placed. At time *T*, a replenishment order with a lot size of Q_1 is placed,

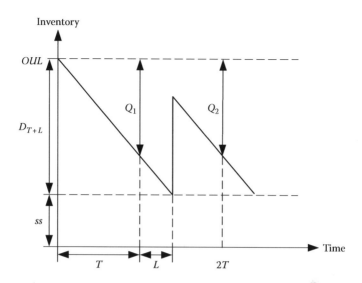

FIGURE 4.7
Inventory profile given a periodic review policy.

which arrives at time $T + L$. At time $2T$, another replenishment order with a lot size of Q_2 is placed. The lot size of each order may vary. The safety inventory is simply the difference between OUL and the expected demand during the time interval $T + L$, as follows:

$$ss = OUL - D_{T+L} \tag{4.23}$$

The CSL given this inventory policy is

$$CSL = F(OUL, D_{T+L}, \sigma_{T+L}) \tag{4.24}$$

From Equation 4.24, we can derive the required OUL to achieve a certain level of CSL, as follows:

$$OUL = F^{-1}(CSL, D_{T+L}, \sigma_{T+L}) \tag{4.25}$$

Substituting $F^{-1}(CSL, D_{T+L}, \sigma_{T+L})$ into Equation 4.23 for OUL, we have

$$ss = F^{-1}(CSL, D_{T+L}, \sigma_{T+L}) - D_{T+L} = F_S^{-1}(CSL) \times \sigma_{T+L} \tag{4.26}$$

Note the similarity between Equations 4.19 and 4.26. There are two factors affecting the required safety inventory, namely, the desired level of CSL and the standard deviation of demand during lead time (replenishment lead time in the case of continuous review and the sum of review lead time and replenishment lead time in the case of periodic review). The average demand during lead time has no effect on the required safety inventory.

Now let us revisit the example in Section 4.3.1; suppose BVE wants to use a periodic review policy with a 4-week review interval while maintaining the same level of *CSL*, what *OUL* should be used? What is the required safety inventory? We have $D_{T+L} = D_W \times (T+L) = 1000 \times (4+2) = 6000$, $\sigma_{T+L} = \sqrt{T+L}\sigma_W = \sqrt{4+2} \times 50 = 122.5$. From Equation 4.25, we have $OUL = F^{-1}(CSL, D_{T+L}, \sigma_{T+L}) = F^{-1}(95\%, 6000, 122.5) \approx 6201$. Under this periodic review policy, BVE reviews its inventory of AAA batteries every 4 weeks and then places an order of $6201 - v$ packs of AAA batteries after the review, where v is the inventory of AAA batteries at the time of review. From Equation 4.23, we can calculate the safety inventory as $ss = OUL - D_{T+L} = 6201 - 6000 = 201$.

The question now is how to determine the review interval under a periodic review policy. This problem is nontrivial because the optimal review interval depends not only on the average yearly demand, inventory holding cost, and fixed ordering cost but also on the demand uncertainty. One simple solution is to first determine the optimal ordering frequency per year, denoted f^*, assuming there is no demand uncertainty. The optimal review interval will be less than or equal to $1/f^*$ year due to the need for carrying safety inventory. The yearly costs under review intervals that are close to $1/f^*$ are then calculated. The review interval that gives the lowest yearly cost is then chosen as the optimal review interval. Referring to the previous example, assume the lot size of 3000 is the *EOQ*, which is 3 weeks' worth of demand. The optimal ordering frequency is thus every 3 weeks, or 17.33 times a year. The review interval should not be more than 3 weeks. We can try review intervals of 3 and 2 weeks. Suppose the review interval of 2 weeks gives a higher cost than the review interval of 3 weeks; we then choose 3 weeks to be the optimal review interval.

Note that if there is no uncertainty associated with the demand, a 3-week periodic review policy will be identical to a continuous review policy with a lot size of 3000. However, the safety inventory required for the 3-week periodical review policy is 184 packs of AAA batteries, which is higher than the 116 packs of safety inventory under the continuous review policy. In general, to maintain the same level of *CSL*, a periodic review policy requires a larger amount of safety inventory than that under a continuous review policy. When deciding which policy to use, one should consider the tradeoff between the increase in the cost of checking inventory and the savings that resulted from safety inventory reduction.

4.3.3 Impact of Demand Correlation

When determining safety inventory, we often need to aggregate demand, either from different time periods or from different sources. Refer to the example in Section 4.3.1, the weekly demand follows a normal distribution and the replenishment lead time is 2 weeks. To determine the demand during lead time, we need to sum up the demands in 2 weeks, that is, aggregate the demands from two different time periods. If a company supplies to several customers, to determine the overall demand within a certain time period, the company needs to sum up the demands from all of its customers. In general, let X_i denote the demand from time period i (or customer i). X_i is normally distributed with a mean μ_i and a standard deviation σ_i. To sum up the demand from n periods (or n customers), we need to know how the demands are correlated. Let ρ_{ij} ($0 \le \rho_{ij} \le 1$) denote the correlation between X_i and X_j, i, $j = 1, 2, \ldots, n$, $i \ne j$. The sum of demands from the n periods (or n customers) follows a normal distribution with mean μ_Σ and standard deviation σ_Σ, where

$$\mu_\Sigma = \sum_{i=1}^{n} \mu_i \quad \text{and} \quad \sigma_\Sigma = \sqrt{\sum_{i=1}^{n} \sigma_i^2 + 2\sum_{i=1}^{n-1}\sum_{j=i+1}^{n} \rho_{ij}\sigma_i\sigma_j} \qquad (4.27)$$

Note that if the demands are independent, that is, $\rho_{ij} = 0$, Equation 4.27 reduces to

$$\mu_\Sigma = \sum_{i=1}^{n} \mu_i \quad \text{and} \quad \sigma_\Sigma = \sqrt{\sum_{i=1}^{n} \sigma_i^2} \qquad (4.28)$$

Independent demand was the assumptions that we made when solving the example problems in Sections 4.3.1 and 4.3.2. Now let us look at the impact of demand correlation. We vary the correlation between the demands in two consecutive weeks from 0 to 1 with an increment of 0.25 and recalculate the required safety inventory for AAA batteries at BVE store under the continuous review policy. The result is shown in Table 4.4. We can see that as the correlation between weekly demand increases,

TABLE 4.4

Impact of Demand Correlation on Safety Inventory

Correlation ρ_{12}	Standard Deviation during Lead Time $\sigma_L = \sqrt{\sum_{i=1}^{2} 50^2 + 2 \times \rho_{12} \times 50 \times 50}$	Safety Inventory $F_S^{-1}(0.95) \times \sigma_L$
0	70.71	116
0.25	79.06	130
0.5	86.60	142
0.75	93.54	154
1	100.00	164

the standard deviation of the demand during lead time also increases and so does the required safety inventory. Note that when the demands between two consecutive weeks are perfectly correlated (i.e., the correlation is 1), the standard deviation of the aggregated 2-week demand is 100, which is exactly twice the weekly demand standard deviation of 50. In fact, when demands are perfectly correlated, the standard deviation of the aggregated demand is simply the sum of the standard deviations of individual demands. If the demands are not perfectly correlated, the standard deviation of the aggregated demand is smaller than the sum of the standard deviations of individual demands. Companies often take advantage of this fact by aggregating inventory from different sources to reduce the required safety inventory while maintaining the same level of *CSL*. We will illustrate this strategy using an example as follows.

Pop Appliances has four store locations in Metropolis. Currently refrigerators are sold in each location and the company offers same day free delivery. Monthly demand in each location follows a normal distribution with the parameters of the distribution shown in Table 4.5. The information of demand correlation is shown in Table 4.6. Pop Appliances reviews its inventory twice a month. The replenishment lead time for refrigerators is half a month. The company aims to achieve a *CSL* of 97%. A supply chain consultant suggests that Pop Appliances carries inventory in only one store location but allows customers to place orders in any locations. Because of the same day free delivery policy, the demand patterns are expected to remain the same under the new arrangement. Determine the reduction in the required safety inventory if the consultant's suggestion is implemented.

The inventory review interval is $T = 0.5$ month and the replenishment lead time is $L = 0.5$ month. We have $T + L = 1$ month. Thus, we need to determine the mean and the standard deviation of demand in 1 month. Under the

TABLE 4.5

Refrigerator Demand Patterns at Pop Electronic

Store Location	1	2	3	4
Mean (μ)	100	80	70	90
Standard deviation (σ)	30	15	20	25

TABLE 4.6

Demand Correlation between Different Store Locations

Store Location	1	2	3	4
1	—	$\rho_{12} = 0.5$	$\rho_{13} = 0.3$	$\rho_{12} = 0.1$
2	—	—	$\rho_{23} = 0.6$	$\rho_{24} = 0.2$
3	—	—	—	$\rho_{34} = 0.4$
4	—	—	—	—

current arrangement, each store location carries its own safety inventory, and the mean and the standard deviation of monthly demand at each location are shown in Table 4.5. Using the data, we can find the safety inventory for store locations 1, 2, 3, and 4 as 56, 28, 38, and 47, respectively. The total safety inventory Pop Appliances needs to carry is 169 units of refrigerator.

The supply chain consultant's proposal is to aggregate the inventory of refrigerators in one location. Therefore, we need to sum up the demands from the four store locations. Using Equation 4.27, we have

$$\mu_\Sigma = \sum_{i=1}^{4} \mu_i = 100 + 80 + 70 + 90 = 340 \text{ and}$$

$$\sigma_\Sigma^2 = \sum_{i=1}^{4} \sigma_i^2 + 2\sum_{i=1}^{3}\sum_{j=i+1}^{4} \rho_{ij}\sigma_i\sigma_j$$

$$= \sigma_1^2 + \sigma_2^2 + \sigma_3^2 + \sigma_4^2 + 2(\rho_{12}\sigma_1\sigma_2 + \rho_{13}\sigma_1\sigma_3 + \rho_{14}\sigma_1\sigma_4 + \rho_{23}\sigma_2\sigma_3$$

$$+ \rho_{24}\sigma_2\sigma_4 + \rho_{34}\sigma_3\sigma_4)$$

$$= 30^2 + 15^2 + 20^2 + 25^2 + 2\times(0.5\times30\times15\times0.3\times30\times20\times0.1\times30\times25$$

$$+ 0.6\times15\times20 + 0.2\times15\times25 + 0.4\times20\times25)$$

$$= 4020$$

Therefore, $\sigma_\Sigma = \sqrt{4020} = 63.40$. Now we can use Equation 4.26 to calculate the safety inventory as $ss = F_S^{-1}(CSL)\times\sigma_\Sigma = F_S^{-1}(0.97)\times63.40 \approx 119$. This is a reduction of 50 units of refrigerators compared to the current arrangement.

4.3.4 Impact of Lead Time Uncertainty

By now we should be clear that the key to calculate the required safety inventory given a certain level of *CSL* is to quantify demand uncertainty during lead time by determining its standard deviation (assuming demand follows a normal distribution). In previous discussion, we only take into account demand uncertainty and assume that the replenishment lead time is constant. In reality, there is uncertainty associated with replenishment lead time as well. A supplier may run into issues with its production facility and thus unable to ship its products on time. Severe weather condition may cause delay in the transportation of goods. When the lead time is uncertain, the uncertainty of demand during lead time will increase. To capture the effect of lead time uncertainty, we assume that the lead time follows a normal distribution with a mean L and a standard deviation σ_E (measured using a certain unit of time). Let μ_D and σ_D denote the mean and standard deviation of demand, respectively, measured using the same unit of time. The standard deviation of demand during lead time, σ_L, is then calculated as (Hadley and Whitin 1963).

$$\sigma_L = \sqrt{L\sigma_D^2 + \mu_D^2 \sigma_E^2} \tag{4.29}$$

With Equation 4.29 we can now take lead time uncertainty into consideration when calculating the required safety inventory given a desired level of *CSL*. Let us revisit the Pop Appliances example in Section 4.3.3 to illustrate how the calculation is done.

Pop Appliances implemented the supply chain consultant's suggestion and now only holds inventory in store location 3. Meanwhile, the consultant noticed that Pop Appliances reviews its refrigerator inventory on the 1st and the 16th of each month, which is not exactly 15 days apart due to varying number of days in each month. In addition, there are variations in the replenishment lead time as well. Based on historical data, it was estimated that the sum of review lead time and replenishment lead time follows a normal distribution with a mean of 30 days and a standard deviation of 3 days. What is the safety inventory required in order to maintain a *CSL* of 97%? If Pop Appliances chooses to hold 119 units of refrigerators based on the calculation done in Section 4.3.3, what is the expected *CSL*?

Our first step is to determine the standard deviation of demand during lead time using Equation 4.29. Note that the time unit used to measure lead time is day. Therefore, we need to determine the mean and the standard deviation of daily customer demand. We have previously calculated the mean and the standard deviation of aggregated monthly demand of refrigerators at Pop Appliances as 340 and 63.40, respectively. To convert monthly demand to daily demand, we assume that there are 30 days in a month and daily demands are independent. Under these assumptions, the mean and the standard deviation of daily demand are $340/30 = 11.33$ and $63.40/\sqrt{30} = 11.58$, respectively. Note that if daily demands are correlated to a certain degree, the standard deviation of daily demand will be smaller than 11.58. Therefore, our assumption of independent daily demand is conservative, that is, it will lead to a higher level of required safety inventory. Using Equation 4.5, we have $\sigma_L = \sqrt{L\sigma_D^2 + \mu_D^2 \sigma_E^2} = \sqrt{30 \times 11.58^2 + 11.33^2 \times 3^2} = 71.96$. The required safety inventory to achieve a *CSL* of 97% is $ss = F_S^{-1}(CSL) \times \sigma_L = F_S^{-1}(0.97) \times 71.96 \approx 135$. Note that $F_S^{-1}(0.97)$ can be quickly calculated in Microsoft Excel by typing "=NORMINV(0.97,0,1)" in a cell and hitting the Enter key.

Now we consider the expected *CSL* if Pop Appliances holds a safety inventory of 119 units of refrigerator. The average lead time is $L=30$ days, and the demand during lead time is $D_L=340$ units of refrigerator. The standard deviation of demand during lead time is $\sigma_L=71.96$. Therefore, $CSL = F(ss + D_L, D_L, \sigma_L) = F(119 + 340, 340, 71.96) = 0.95$. This example shows that with the same level of safety inventory, lead time uncertainty will reduce the expected *CSL*.

4.4 Optimal Product Availability

The level of product availability, commonly measured using *CSL*, is a primary criterion to evaluate the responsiveness of a company. A company can use a high *CSL* to improve its responsiveness and attract customers, thus increasing its revenue. However, to achieve a high *CSL* requires a high level of safety inventory, which increases inventory holding cost. Therefore, there is a need to determine the optimal level of product availability.

Two key factors that influence the optimal level of product availability are (1) the cost of overstocking the product and (2) the cost of understocking the product. The overstocking cost, denoted as C_o, is the cost of holding a unit of an unsold product at the end of the selling season. The understocking cost, denoted as C_u, is the marginal loss incurred for the lost sale of a unit of product due to the lack of inventory. Assume that the demand is a random variable with density function $f(x)$ and cumulative distribution function $F(x)$. Let Q be the number of units of inventory on hand. If $Q < x$, then we incur understocking cost of $(x - Q)C_u$. If $Q > x$, then we incur overstocking cost of $(Q - x)C_u$. The expected cost $C(Q)$ is calculated as follows:

$$C(Q) = \int_Q^\infty (x - Q)C_u f(x)dx + \int_0^Q (Q - x)C_o f(x)dx \tag{4.30}$$

To find the value of Q that minimizes the expected cost, denoted Q^*, we take the derivative of $C(Q)$ against Q and set it to zero. We have

$$\frac{dC(Q)}{dQ} = -C_u \int_Q^\infty f(x)dx + C_o \int_0^Q f(x)dx = -C_u(1 - F(Q)) + C_o F(Q) = 0 \tag{4.31}$$

Therefore,

$$F(Q^*) = \frac{C_u}{C_u + C_o} \tag{4.32}$$

Note that $F(Q^*) = P(x \le Q^*)$. Refer to Equation 4.17. Q^* corresponds to the optimal *ROP* and $F(Q^*)$ corresponds to the optimal *CSL*. Therefore, given C_o and C_u, the optimal *CSL* can be calculated as

$$CSL^* = \frac{C_u}{C_u + C_o} \tag{4.33}$$

Now the question is how to determine overstocking and understocking cost. For continuously stocked products, the overstocking cost is the inventory holding cost during a replenishment cycle. Let D be the annual demand, H be the annual inventory holding cost, and Q be the replenishment lot size. There will be D/Q replenishment cycles each year. Therefore, $C_o = H/(D/Q) = QH/D$. If demand during stockout is lost, the understocking cost C_u is the lost profit. In this case, we have

$$CSL^* = \frac{C_u}{C_u + C_o} = \frac{C_u}{C_u + QH/D} = \frac{DC_u}{DC_u + QH} \qquad (4.34)$$

If demand during stockout is backlogged, let C_d denote the backlog cost, e.g., a discount of C_d is offered to keep the customer demand. Because a backlogged unit is delivered to the customer when the new replenishment order arrives, no inventory holding cost is incurred for the backlogged unit. Under this circumstance, the understocking cost C_u is $C_d - (QH/D)$. Therefore, we have

$$CSL^* = \frac{C_u}{C_u + C_o} = \frac{C_d - QH/D}{C_d - QH/D + QH/D} = \frac{C_d - QH/D}{C_d} = 1 - \frac{QH}{DC_d} \qquad (4.35)$$

Therefore, for continuously stocked products, we can use either Equation 4.34 or 4.35 to calculate the optimal CSL, depending on whether demand during stockout is lost or backlogged. After quantifying demand uncertainty during lead time by determining its standard deviation, we can then calculate the desired safety inventory to minimize expected cost. Unsold product units at the end of each replenishment cycle will be carried over to the next cycle.

For seasonal products with a single order in a season, we will need to determine the optimal order quantity instead of the required safety inventory. Note that for such products, unsold units must be disposed of at the end of the season. Assume a unit of leftover product has a salvage value of s, its cost is c, and its sales price is p. We have $C_u = p - c$ and $C_o = c - s$. The optimal CSL is

$$CSL^* = \frac{C_u}{C_u + C_o} = \frac{p - c}{(p - c) + (c - s)} = \frac{p - c}{p - s} \qquad (4.36)$$

Assume demand during the season follows a normal distribution with a mean μ and a standard deviation σ, the optimal order quantity is

$$Q^* = F^{-1}(CSL^*, \mu, \sigma) \qquad (4.37)$$

In the event that a manufacturer offers quantity discount, we will have different costs at different price breaks. To determine the optimal order quantity, we can use procedures similar to that presented in Section 4.2.4.

The only difference is that instead of minimizing the inventory cost we should maximize the expected profit of ordering Q units of product. Note that if Q units are ordered and the demand $x < Q$, then x units are sold that produces a profit of $(p - c)x$. There will be $Q - x$ units of unsold product that results in a loss of $(c - s)(Q - x)$. Therefore, the overall profit is $(p - c)x - (c - s)$ $(Q - x) = (p - s)x - (c - s)Q$. On the other hand, if the demand $x \geq Q$, then all Q units are sold, which produces an overall profit of $(p - c)Q$. Therefore, the expected profit for ordering Q units of product is

$$E(Q) = \int_{-\infty}^{Q} [(p-s)x - (c-s)Q]f(x)dx + \int_{Q}^{\infty} (p-c)Qf(x)dx$$

$$= (p-s)\mu F(Q) - (p-s)\sigma^2 f(Q) - (c-s)QF(Q) + (p-c)Q[1-F(Q)] \qquad (4.38)$$

The expected overstock or understock by ordering Q units can also be determined. Overstock occurs when the demand $x < Q$. In this case, the expected overstock is

$$\int_{-\infty}^{x} (Q-x)f(x) = QF(Q) - \left[\mu F(Q) - \sigma^2 f(Q)\right]$$

$$= (Q-\mu)F(Q) + \sigma^2 f(Q) \qquad (4.39)$$

Understock occurs when the demand $x > Q$. In this case, the expected understock is

$$\int_{Q}^{\infty} (x-Q)f(x)dx = \mu[1 - F(Q)] + \sigma^2 f(Q) - Q[1 - f(Q)]$$

$$= (\mu - Q)[1 - F(Q)] + \sigma^2 f(Q) \qquad (4.40)$$

We now use an example to illustrate how to determine the optimal order quantity for seasonal products taking into account quantity discount.

Consider the sales of digital TV converters at BVE store. With the conversion of analog to digital TV, BVE determines that there is an opportunity to sell digital TV converters over the next 6 months. Any remaining converters at the end of the selling season will be disposed of with no salvage value. The demand is estimated to follow a normal distribution with a mean of 5000 and a standard deviation of 500. The converter has a sales price of $75. The converter manufacturer plans to charge $30 per unit if the order is less than 6000 units and $25 per unit if the order is at least 6000 units. How many converters should BVE order?

The manufacturer offers all quantity discount for its digital TV converters. Therefore, we should follow the corresponding procedure and start from the lowest purchasing cost of $c = 25$. We have the sales price $p = 75$ and the salvage value $s = 0$. Using Equation 4.36, we find that $CSL^* = (p-c)/(p-s) = (75-25)/75 = 0.67$ and $Q^* = F^{-1}(CSL^*, \mu, \sigma) = F^{-1}(0.67, 5000, 500) = 5215$ units. Because $Q^* < 6000$, we need to increase the order quantity to 6000 in order to qualify for the discounted purchasing cost of \$25 per unit. Using Equation 4.38, we can calculate the expected profit for ordering 6000 units of digital TV converter as $E(6000) = (75-0) \times 5000 \times F(6000, 5000, 500) - (75-0) \times 500^2 \times f(6000, 5000, 500) - (25-0) \times 6000 \times F(6000, 5000, 500) + (75-25) \times$

$6000 \times [1 - F(6000, 5000, 500)] \approx \$224,682$. Note that $F(6000, 5000, 500)$ and $f(6000, 5000, 500)$ can be quickly calculated in Microsoft Excel by typing "=NORMDIST(6000,5000,500,1)" and "= NORMDIST(6000,5000,500,0)" in a cell, respectively, and hitting the Enter key.

Now we evaluate the next lowest purchasing cost of $c = 30$. We have $CSL^* = 0.6$ and $Q^* = 5127$, which qualifies for the purchasing cost of \$30 per unit. We then calculate the expected profit of ordering 5127 units of digital TV converter as $E(5127) = \$210,512$. Because the expected profit when ordering 6000 units is higher than that when ordering 5127 units, BVE should order 6000 units of digital TV converters.

4.5 Case Studies

Case Study 4.5.1

Weekly demand of laptop computers in BVE store is independent and normally distributed with a mean of 1000 and a standard deviation of 300. The inventory holding cost is \$152 per laptop per year. If the laptop is out of stock, the store offers \$100 discount per laptop to keep the customer. Replenishment orders are handled by Interstate Shipping, which provides two shipping options. Regular shipping costs \$6000 per order and takes 2 weeks for delivery, whereas expedited shipping costs \$7000 per order and takes 1 week for delivery. The laptop inventory at BVE is monitored continuously. How do we manage laptop inventory at BVE?

The average weekly demand is $D_W = 1000$ and the weekly standard deviation is $\sigma_W = 300$. Therefore, the yearly demand is $D = 52 \times D_W = 52,000$. Demand during stockout is backlogged with discount $C_d = 100$. The inventory holding cost is $H = 154$. Let us investigate regular shipping first. We need to determine both the cycle inventory and the safety inventory. For cycle inventory, we need to find the EOQ, which is $Q^* = \sqrt{2DO/H} = \sqrt{(2 \times 52,000 \times 6,000)/152} \approx 2,026$. Therefore, the cycle inventory is $Q^*/2 = 1013$. The annual ordering cost is $DO/Q^* = (52,000 \times 6,000)/2,026 \approx \$153,998$. For safety inventory, we need to determine the optimal CSL first. Because demand during stockout is backlogged, we use Equation 4.35 and have

$CSL^* = 1 - (Q^* H/DC_d) = 1 - ((2,026 \times 152)/(52,000 \times 100)) = 0.9408$. Because the replenishment lead time is 2 weeks and weekly demands are independent, the reorder point is $ROP = F^{-1}(CSL^*, D_L, \sigma_L) = F^{-1}(0.9408, 2 \times 1000, \sqrt{2} \times 300) \approx 2662$. The safety inventory is $ss = ROP - D_L = 2662 - 2 \times 1000 = 662$. Therefore, the average annual inventory is $(Q^*/2) + ss = 1013 + 662 = 1675$. The annual inventory holding cost is thus $1675 \times 152 = \$254{,}600$. The total annual inventory cost is $\$153{,}998 + \$254{,}600 = \$408{,}598$.

Similarly, we investigate the use of expedited shipping. The EOQ is $Q^* = \sqrt{((2 \times 52,000 \times 7,000)/152)} \approx 2{,}188$. The cycle inventory is 1094. The annual ordering cost is $DO/Q^* = (52,000 \times 7,000)/2,188 \approx \$166{,}362$. The optimal CSL is $CSL^* = 1 - ((2,188 \times 152)/(52,000 \times 100)) = 0.9360$. The reorder point is $ROP = F^{-1}(0.9360, 1000, 300) \approx 1457$. The safety inventory is $ss = 1457 - 1 \times 1000 = 457$. The average annual inventory is $1094 + 457 = 1551$. The annual inventory holding cost is $1551 \times 152 = \$235{,}752$. The total annual inventory cost is $\$166{,}362 + \$235{,}752 = \$402{,}114$.

Our analysis showed that the total annual inventory cost using the expedited shipping option is \$6484 lower than that using the regular shipping option. This is because using the expedited shipping option allows us to hold less safety inventory. The savings resulted from lower safety inventory exceeds the additional cost of using expedited shipping. Therefore, BVE store should choose the expedited shipping option and order 2188 units of laptops whenever the inventory falls to 1457 units. In general, the required safety inventory is reduced when the replenishment lead time is shortened. This will lead to lower inventory holding cost and thus improve the profitability of a company.

Case Study 4.5.2

Ultra Corp. produces two different models of laptop computers, Ultra Pro and Ultra Home. The weekly demand for Ultra Pro is independent and normally distributed with a mean of 800 and a standard deviation of 400. The weekly demand for Ultra Home is also independent and normally distributed with a mean of 100 and a standard deviation of 250. Currently Ultra Pro is equipped with LG8555 battery, and Ultra Home is equipped with LG8111 battery. Both LG8555 and LG8111 batteries are produced by Longey, so they are always ordered jointly. The replenishment lead time is 4 weeks. LG8555 has a slightly longer battery life than that of LG8111. Its cost is \$80 per unit, whereas the cost of LG8111 is \$78. The annual inventory holding cost for one unit of battery is 20% of its cost. The company aims to achieve a CSL of 99%. The demands for Ultra Pro and Ultra Home are independent. Mary, the supply chain manager at Ultra Corp., suggested that the company also equip Ultra Home with LG8555 batteries to simplify inventory management. Can her suggestion save money for the company?

LG8555 batteries are more expensive than LG8111 batteries, so Mary's suggestion seems counterintuitive. However, using a common component for both laptop models could lower the required safety inventory, which results in lower inventory holding cost. If the reduction in inventory holding cost is more than the increase in component cost, it would make sense to use a better and slightly more expensive component for both laptop models. This way, the company can provide better value to its customers while at the same time reduce its cost.

We first analyze the required safety inventory when using two different battery models. The safety inventory required for LG8555 is $ss = F_S^{-1}(CSL) \times \sigma_L = F_S^{-1}(0.99) \times \sqrt{4} \times 400 \approx 1861$ units. The annual cost for holding the safety inventory is $1,861 \times 80 \times 20\% = \$29,776$. Similarly, the safety inventory required for LG8111 is 1,163 units with an annual holding cost of \$18,143. Therefore, the current policy of using two different battery models incurs a cost of \$47,919 for holding the required safety inventory.

If only LG8555 battery is used, the weekly demand is normally distributed with a mean of $800 + 100 = 900$ units and a standard deviation of $\sqrt{400^2 + 250^2} = 471.70$ units. The required safety inventory is $ss = F_S^{-1}(0.99) \times \sqrt{4} \times 471.70 \approx 2195$ units. The annual cost for holding the safety inventory is \$35,120. Therefore, we realize an annual saving of \$12,799. The annual increase in component cost is $100 \times 52 \times (80 - 78) = \$10,400$. The cost reduction in holding the required safety inventory is more than the increase in component cost. Therefore, Mary's suggestion can indeed save money for Ultra Corp.

Case Study 4.5.3

Jackson Heating and Cooling (JHC) strives to provide the best service to its customer. Its main business has been the installation of air conditioners during the months from June to September. The demand fluctuates significantly from day to day, with an average of 10 units per day and a standard deviation of 20 units per day. JHC purchases air conditioners from Metro Air Distributors (MAD) and maintains a warehouse to store its inventory. The replenishment lead time follows a normal distribution with a mean of 5 days and a standard deviation of 5 days. JHC implements a continuous review policy with a *CSL* of 0.99. Therefore, it holds a safety inventory of 156 units. JHC generally reduces its order toward the end of September to clear out the safety inventory. However, because of the high demand fluctuations and its goal to provide the best customer service possible, JHC usually has a large number of leftover air conditioners that it has to carry over to the next year. Recently, JHC has seen steady increase in the sales of its high-efficiency gas furnace during winter time. Carrying leftover air conditioners to next summer creates a storage problem for the gas furnaces in winter. To solve the storage problem, JHC estimated that it needs to reduce the safety inventory of its air conditioners by 30%. This can be achieved by reducing the standard deviation of replenishment lead time from 5 days to 1 day, while maintaining the same level of *CSL*. If this reduction in safety inventory can be done, JHC estimated that it will save \$40,000 per year.

Dale Humpert, a retired executive from MAD, is now a member of JHC's board of directors. He knows that the fluctuation of replenishment lead time that JHC experienced is mainly due to the fact that MAD maintains a *CSL* of only 90%. If the *CSL* at MAD is increased to 99%, then the standard deviation of replenishment lead time can be reduced to 1 day, meeting JHC's need. He also knows that MAD has at least 10 customers similar to JHC. These customer demands are independent. Dale has an idea of organizing these customers to form a strategic partnership with MAD. MAD will increase its *CSL* to 99% to improve its performance in on-time delivery; so the customers can carry less safety inventory while maintaining the same level of *CSL*. In return, these customers will pay MAD a certain percent of their annual savings resulted from the reduction in safety inventory. MAD's replenishment lead time is exactly 14 days. Its inventory holding cost is $700 per unit of air conditioner per year. Is Dale's idea feasible?

Safety inventory is held at different stages of a supply chain. With proper coordination it is possible to reduce the total safety inventory held in a supply chain while maintaining the overall customer service performance of the supply chain. This is the basis for Dale's idea. We know that JHC will save $40,000 annually due to safety inventory reduction if MAD increases its safety inventory to achieve a *CSL* of 99%. With 10 customers similar to JHC, the total annual savings will be $400,000. Therefore, if the cost for MAD to improve its *CSL* to 99% is considerably lower than $400,000, then Dale has a feasible idea.

We first determine the aggregated demand of the 10 customers at MAD. The aggregated daily demand follows a normal distribution with a mean of $10 \times 10 = 100$ and a standard deviation of $\sqrt{10} \times 20 = 63.25$. During MAD's replenishment lead time, the standard deviation of its customer demand is $\sqrt{14} \times 63.25 = 236.66$. Currently MAD maintains a *CSL* of 90%, which requires a safety inventory of $F_S^{-1}(0.90) \times 236.66 \approx 303$ units of air conditioners. To achieve a *CSL* of 99%, MAD needs to increase its safety inventory to $F_S^{-1}(0.99) \times 236.66 \approx 551$. The increase in MAD's annual safety inventory holding cost is $(551 - 303) \times 700 = \$173,600$. This is less than half of the total annual savings that the 10 customers achieved. Dale's idea is certainly feasible. Note that currently the total safety inventory in the supply chain is $303 + 10 \times 156 = 1863$ units of air conditioners. If Dale's idea is realized, each customer would only need to hold 107 units of safety inventory. The total safety inventory in the supply chain is thus $551 + 10 \times 107 = 1621$ units of air conditioners. As a result, the supply chain can reduce its total safety inventory by 215 units of air conditioners. The cost savings can then be shared among MAD and its customers.

Case Study 4.5.4

Lily Sharp started a small business selling premium European chocolate 5 years ago. Her business grew steadily in the first 3 years. During the past 2 years, the sales leveled off. With limited budget, Lily tried advertising and occasional price discount. However, these efforts did not lead to meaningful improvement in sales. The weekly demand of chocolate in Lily's store now roughly follows a normal distribution with a mean of 300 lb and a standard deviation of 100 lb. Lily has been ordering chocolate from Vogel Trading Corp., which offers incremental discount with three price breaks. For quantity under 3000 lb, the price is \$20 per lb. For quantity between 3,000 and 10,000 lb, the price is \$18 per lb. For quantity above 10,000 lb, the price is \$15 per lb. Lily estimated that her capital cost is 10% a year and storage cost is \$2.5 per lb of chocolate per year. She is a very organized person and always makes a quick but accurate estimate of her inventory at the end of each day. She uses a logistics company to handle the ordering and delivery of chocolate. The logistics company charges a fixed \$500 fee per order for up to 3000 lb, plus \$1 per lb for any additional chocolate ordered. Because Vogel Trading Corp. is a distributor that imports chocolate from Europe and Lily is not a major customer, once in a while Lily's order gets delayed when the distributor experiences stockout. It was estimated that the replenishment lead time follows a normal distribution with a mean of 1 week and a standard deviation of 2 weeks. To deal with this lead time uncertainty, Lily maintains a safety inventory of 900 lb of chocolate, equals to 3 weeks of demand. She places an order of 1800 lb of chocolate (6 weeks' worth of demand) whenever her inventory falls below 1200 lb.

Lily has been exploring different ways to expand her business. She found out that if a customer orders more than 50,000 lb of chocolate per year from Vogel Trading Corp., the customer is treated as a valued customer with guaranteed on-time delivery. In other words, if Lily qualifies as a valued customer, her replenishment lead time will be exactly 1 week. In addition, she identified four local stores who have comparable chocolate sales and also order from Vogel Trading Corp. These stores' order quantity is around 2000 lb of chocolate per order, which means their cost base is \$20 per lb. The demands among these stores are independent. Lily has come up with a new business plan as follows. She would offer to become the chocolate supplier of the four stores, with a cost of \$20 per lb. She would promise 2-h delivery with a *CSL* of 99%, where the delivery cost is paid by the customer, that is, the stores. She is confident that the stores will accept her offer because of the following cost benefits: (1) the stores can drastically reduce, if not totally eliminate, their safety inventory because replenishment can be completed within 2 h; (2) the cost of local delivery will be lower than the cost of ordering and receiving shipment from Vogel Trading Corp. Will Lily's new business plan generate profit for her?

Let us look at Lily's current inventory management practice. First, we study the implied *CSL* under Lily's safety inventory policy. The average and the standard deviation of replenishment lead time are $L=1$ week and $\sigma_E=2$ weeks, respectively. The average and the standard deviation of weekly demand are $\mu_D=300$ lb and $\sigma_D=100$ lb, respectively. The average and the standard deviation of demand during lead time are thus $D_L=300$ lb and $\sigma_L=\sqrt{1\times100^2+300^2\times2^2}=608.28$ lb. Lily's reorder point is $ROP=1200$. From Equation 4.18, we have $CSL=F(ROP,D_L,\sigma_L)=F(1200,300,608.28)=93\%$.

TABLE 4.7

Calculation of Optimal Order Quantity under Current Demand

i	q_i	c_i	Q_i^*	$P(Q_i^*)$	$TC(Q_i^*)$
0	0	—	—	—	—
1	3,000	20	1,862	$37,240	$324,568
2	10,000	19	4,982	$97,658	$329,580
3	∞	16	15,966	$288,456	$351,093

Now we look at Lily's order lot size. We have $D = 52 \times 300 = 15,600$, $O = 500$, $W = 2.5$, and $\alpha = 10\%$. Note that if the order quantity exceeds 3000 lb, Lily has to pay a surcharge of $1 per lb to the logistics company for the additional quantity. This is equivalent to the increase of her cost of chocolate by $1 per lb for the marginal quantity. We can take this additional cost into account by revising the price schedule as follows. For order quantity up to 3000 lb, the price is still $20 per lb. For order quantity between 3,000 and 10,000 lb, the price is $19 per lb. For order quantity above 10,000 lb, the price is $16 per lb. We then use the procedure presented in Section 4.2.4 to determine the optimal order quantity, as shown in Table 4.7. The optimal order quantity turns out to be 1862 lb of chocolate. This is very close to Lily's lot size of 1800 lb, indicating that Lily is doing a good job in managing her inventory. Lily's annual logistics cost is $(15,600 \div 1,800) \times 500 = \$4,333$. Her annual inventory holding cost is $((1800/2) + 900) \times (0.1 \times 20 + 2.5) = \$8100$. The total annual inventory cost is $4,333 + 8,100 = \$12,433$.

By becoming a supplier to the four stores, Lily will quintuple the demand for chocolate in her business. Because the demands in these four stores are comparable to that in Lily's business, we assume the weekly demands in these stores all follow a normal distribution with a mean of 300 lb and a standard deviation of 100 lb of chocolate. The total annual demand will be $5 \times 52 \times 300 = 78,000$ lb of chocolate, which will qualify Lily as a valued customer of Vogel Trading Corp and earn her a guaranteed replenishment lead time of exactly 1 week. The aggregated weekly demand follows a normal distribution with a mean $D_L = 5 \times 300 = 1500$ lb and a standard deviation $\sigma_L = \sqrt{5 \times 100^2} = 223.61$ lb of chocolate. To achieve CSL of 99%, the required safety inventory is $ss = F_S^{-1}(CSL) \times \sigma_L = F_S^{-1}(0.99) \times 223.61 \approx 520$ lb of chocolate. This is lower than the 900 lb of safety inventory Lily currently holds.

Lily would also need to determine the optimal order quantity now that her annual demand increases fivefold. With the revised price schedule (taking into account the logistics company's surcharge for order quantity above 3000 lb) and the new annual demand of $D = 78,000$ lb of chocolate (the fixed ordering cost, storage cost, and capital cost remain the same), the optimal order quantity turns out to be 35,702 lb, as shown in Table 4.8. Adding the surcharge by the logistics company to Lily's chocolate purchasing cost, we can calculate the average

TABLE 4.8

Calculation of Optimal Order Quantity
under Expected Demand

i	q_i	c_i	Q_i^*	$P(Q_i^*)$	$TC(Q_i^*)$
0	0	—	—	—	—
1	3,000	20	4,163	$83,260	$1,588,102
2	10,000	19	11,140	$213,660	$1,555,822
3	∞	16	35,702	$604,232	$1,470,867

cost of chocolate as $604,232 \div 35,702 = \$16.92$. Therefore, Lily's annual logistics cost (excluding the surcharge because it has been added to the purchase cost) is now $(78,000 \div 35,702) \times 500 = \$1,092$. Her annual inventory holding cost is $((35,702/2) + 520) \times (0.1 \times 16.92 + 2.5) = \$77,011$. The expected total annual inventory cost is $1,092 + 77,011 = \$78,103$, an increase of $65,670 over the current total annual inventory cost. On the other hand, Lily has lowered the cost base of her chocolate from $20 to $16.92 per lb. With the expected annual sales of 78,000 lb of chocolate, Lily will increase her revenue by $78,000 \times (20 - 16.92) = \$240,240$. This will produce an additional profit of $\$240,240 - \$65,670 = \$174,570$. Lily's new business plan is indeed profitable. In fact, she can supply chocolate to the four stores at a price of $19 per lb (to further improve the appeal of her business proposal) and still make a good profit.

Problem: Business Attire Store

Business Attire Store (BAS) sells both men's and women's business suits. A men's suit costs $100 and is sold for $200. Its daily demand is independent and normally distributed with a mean of 120 and a standard deviation of 100. The inventory holding cost for a men's suit is $40 a year. A women's suit costs $120 and is sold for $250. Its daily demand is independent and normally distributed with a mean of 80 and a standard deviation of 60. The inventory holding cost for a women's suit is $45 a year.

BAS opens 365 days a year. Demand during stock out is lost. It currently adopts a periodic review policy for inventory management. It reviews its inventory every 30 days. Employees must work overtime when checking the inventory. Therefore, BAS incurs a cost of $1000 every time it reviews its inventory; of which 60% is allocated to counting men's suits and 40% to counting women's suits. After reviewing the inventory, BAS orders suits from its suppliers. It takes 3 days for the men's suit supplier to get the order ready, whereas the women's suit supplier takes 5 days to get the order ready. BAS uses its own truck to pick up the order overnight. A trip to the men's suit supplier costs $1500, whereas a trip to the women's suit supplier costs $1700.

If the trips to the men's and women's suit suppliers are combined, the cost is $3000. Currently, BAS waits for 5 days after placing its order and then sends the truck to both suppliers to pick up the orders.

You are hired as a consultant to review BAS's inventory management practice. You are required to answer the following questions:

- Given the current inventory management practice, how much safety inventory should BAS carry?
- Should BAS change the frequency of its periodic review? Should it always review and pick up men's and women's suits at the same time?
- BAS is considering the use of an inventory monitoring system to provide real-time inventory counts. After this system is implemented, a continuous review policy will be adopted and men's and women's suits will be picked up separately. The system costs $500,000 to install and $5,000 per year to maintain. It has an estimated useful life of 25 years. Should BAS use the inventory monitoring system?

Exercises

4.1 Quickturn Tool and Die produces a variety of custom fixtures on a general purpose machine. Products are stored in a warehouse before shipment. The best selling product is three-jaw chucks, with a demand of 64,000 per year. Every time the machine is set up to produce three-jaw chucks, a cost of $1000 is incurred. Each three-jaw chuck costs $40. The warehouse has a capacity to store 100,000 three-jaw chucks and its operating cost is $50,000 per month. The warehouse is also used to store other products, so it is fully utilized throughout the entire year. Assume the interest rate is 5%. To minimize cost, how often should three-jaw chucks be produced and in what quantity?

4.2 Pegasus Motors Corp. has its engine assembly plant in Milwaukee and its motorcycle assembly plant in Pittsburgh. Engines are transported using trucks from Roadway Inc., which costs $4000 per trip. The motorcycle plant opens 50 weeks a year and 5 days a week. Each business day the plant produces 800 motorcycles. The holding cost for each engine is $100 per year.

(a) How often should Pegasus transport engines and in what quantity?

(b) Supercheap Inc. offered to transport engines for Pegasus with the following pricing policy: for each trip a fixed $1000 is charged; if the amount of engines transported is more than 2500, then each additional engine is charged an extra $3. Will Pegasus save money by using Supercheap Inc.?

4.3 Annual demand of vitamins is 100,000 bottles at the Nutrition Store (NS). Currently NS purchases vitamins from General Health (GH) for $10 (capital cost) a bottle with a lot size of 10,000. The fixed ordering cost is $1000. The inventory holding cost is 20% of the capital cost. GH's production cost is $5 per bottle of vitamins. The inventory holding cost is $1 per bottle per year. GH incurs an order filling cost of $3000 each time it ships vitamins to NS.

(a) What is the annual inventory cost (holding + ordering) at NS?

(b) What is the annual inventory cost (holding + order filling) at GH?

(c) If GH offers a price of $9.94 for order quantity of 20,000 and above and NS changes its lot size to 20,000, what is the saving in total annual inventory cost at the two companies?

4.4 Daily demand of car batteries in Best Auto Mart (BAM) is normally distributed with a mean of 15 and a standard deviation of 10. BAM opens 365 days a year. Its replenishment lead time follows a normal distribution with a mean of 5 days and a standard deviation of 2 days. The inventory holding cost at BAM is $17 per battery per year. The fixed ordering cost is $315 per order. BAM targets a cycle service level of 95%. BAM uses a periodic review policy. How often should BAM review its inventory? What OUL should be used?

4.5 Weekly demand of packs of AAA batteries in BVE store is normally distributed with a mean of 1000 and a standard deviation of 50. The fixed ordering cost is $130 per order. The inventory holding cost is $1.5 per pack of batteries per year. The replenishment lead time is 2 weeks. The company uses a continuous review policy with an order lot size of 3000. BVE makes a profit of $1.7 per pack of batteries. Currently demand during stockout is lost.

(a) What is the optimal CSL? What is the level of safety inventory that BVE needs to carry?

(b) BVE is considering a plan to backlog customer demand and mail the batteries to customers when the inventory is replenished. It does not offer any discount but the mailing cost is estimated to be $1 per pack of batteries. What is the optimal CSL under this plan?

References

Chopra, S. and P. Meindl. 2003. *Supply Chain Management: Strategy, Planning, and Operation*, 2nd edn. Upper Saddle River, NJ: Pearson Prentice Hall.

Hadley, G. and T. M. Whitin. 1963. *Analysis of Inventory Systems*. Englewood Cliffs, NJ: Prentice-Hall.

Harris, F. W. 1913. How many parts to make at once. *Factory: The Magazine of Management* **10**: 135–136, 152.

5

Moving Products across Supply Chain: Distribution Network Design and Transportation Decision Making

5.1 Overview

In a supply chain, raw materials and components are moved from suppliers to manufacturers and finished products are moved from manufacturers to consumers. The movement of goods in a supply chain is referred to as *distribution*. Distribution is an important factor in determining the profitability of a supply chain. It affects both the supply chain costs and customer experience. Costs associated with a distribution network are summarized as follows:

- *Facility costs*: Facilities include production plants and warehouses. A supply chain needs to have a sufficient number of facilities to meet customer demand. However, as the number of facilities increases, facility costs also increase.

- *Transportation costs*: Transportation costs include *inbound transportation costs*, which are the costs associated with bringing goods to a facility, and *outbound transportation costs*, which are the costs associated with sending goods out of a facility. Because inbound lot size is generally larger than outbound lot size, inbound transportation cost per unit is lower than outbound transportation cost per unit. Increasing the number of warehouses decreases the travel distance to customers and thus decreases outbound transportation costs. As long as economies of scale is maintained for inbound transportation, increasing the number of facilities decreases total transportation costs.

- *Inventory costs*: Inventory is held at each facility in a supply chain. As the number of facilities in a supply chain increases, the inventory costs also increase.

The sum of these costs is referred to as *total logistics costs*. As the number of facilities increases, total logistics costs first decrease and then increase. Therefore, a company should find an appropriate number of facilities to minimize total logistics costs.

When designing a distribution network, a company needs to consider customer service. After all, a company's revenue is generated from its customers. It must provide satisfactory customer service in order to maintain market share. Customer service measures that are affected by the distribution network are summarized as follows:

- *Delivery lead time*, which is the time it takes for a customer to receive an order.
- *Product variety*, which is the number of different products or product configurations that are offered.
- *Product availability*, which is the probability that a product is in stock when a customer order arrives.
- *Order tracking*, which is the ability of customers to track their orders from placement to delivery.
- *Product return*, which is the ability of customers to return the merchandise they purchased when not satisfied.

A company must understand its customers in order to design an appropriate distribution network. If its customers can tolerate a longer delivery lead time, a company can design a distribution network with only few centrally located facilities to reduce facility and inventory costs. On the other hand, if its customers want a short delivery lead time, then a company must have many facilities located near the customers.

In general, there are six types of design options for a distribution network, summarized as follows (Chopra and Meindl 2010):

- *Manufacturer storage with direct shipping*. Information flows from the customer, via the retailer, to the manufacturer. Products are shipped directly from the manufacturer to the end customers. The retailer does not carry inventory. Instead, inventory is centralized at the manufacturer. The manufacturer can then aggregate demand across all retailers. As a result, the supply chain is able to provide a high level of product availability with lower levels of inventory. This is especially useful for high-value, low-demand items with high-demand uncertainty. This design also makes it possible to postpone product customization until after a customer order is received, which further reduces inventory by aggregating to the component level. In addition, fewer facilities are required, which reduces facility costs. However, transportation costs are high and delivery lead time tends to be long because the travel distances to the end customers are longer. Significant investment in information infrastructure is required to integrate the manufacturer and the retailers.

- *Manufacturer storage with direct shipping with in-transit merge.* If a retailer offers products from different manufacturers, it can use in-transit merge to combine shipments from different manufacturers so the customer receives a single delivery. This reduces transportation costs but increases facility costs and may slightly increase delivery lead time. The requirement in information infrastructure is also higher due to the need of merging shipments.

- *Distributor storage with carrier delivery.* Inventory is held by distributors or retailers in intermediate warehouses instead of held by manufacturers. Package carriers are used to transport products from the intermediate warehouses to the end customer. Under this option, inventory costs are higher than when inventory is held by manufacturers, because the ability to aggregate demand is lower. This option is suitable for products with medium or high demand. Facility costs are higher, but transportation costs are lower because economies of scale can be achieved for inbound transportation to the warehouses. Delivery lead time is shorter compared to manufacturer storage because the entire order is aggregated at the warehouse. Compared to that with manufacturer storage, the required information infrastructure is less complicated.

- *Distributor storage with last-mile delivery.* Instead of using carrier delivery, the distributor/retailer delivers products to customers under this option. The warehouses must be close to the customers, which requires more facilities and thus higher facility costs. Inventory aggregation level is low, so this option is more suitable for fast-moving products for which aggregation does not lead to a significant decrease of inventory. The transportation costs are higher compared to that with carrier delivery due to the lack of economies of scale. On the other hand, delivery lead time is shorter than that with carrier delivery. The level of customer satisfaction is very high under this option.

- *Manufacturer or distributor storage with customer pickup.* Under this option, inventory is stored at the manufacturer or distribution warehouses. Customers place orders and then travel to designated pickup locations to collect the merchandise. Inventory costs can be kept low and the transportation costs are low. Facility costs will increase if new pickup locations need to be built. The ability to track order is critical under this option. The complexity of the required information infrastructure is high.

- *Retail storage with customer pickup.* This is the most traditional distribution option where inventory resides locally at retail stores. The inventory holding costs and facility costs are the highest, but the transportation costs are the lowest because of economies of scale.

Each of these distribution network design options has its own strengths and weaknesses. When designing a distribution network, a company needs to consider product characteristics and the performance requirements of the supply chain. Most companies are best served by a combination of delivery networks. The combination used depends on product characteristics and a company's target customers.

Transportation enables the smooth operation of a distribution network. It moves inbound materials from suppliers to original equipment manufacturers (OEMs), repositions inventory among manufacturing plants and distribution centers, and delivers products to customers. In fact, it is an important social function that has consistently accounted for about 10% of the gross domestic product (GDP) of the United States in the past 20 years, according to the Bureau of Transportation Statistics (http://www.bts.gov/publications/national_transportation_statistics). As previously mentioned, transportation cost is an important part of the costs incurred within a distribution network. In most supply chains, especially global ones, appropriate use of transportation is an important driver to reduce cost and to develop competitive advantages. Therefore, we need to understand different transportation modes and their characteristics. The Bureau of Transportation Statistics classifies transportation modes into six categories, namely, (1) air, (2) highway, (3) railroad, (4) transit, (5) waterborne, and (6) pipeline. Transit refers to public transportation where a large number of passengers travel from one place to another. It is rarely used as a transportation mode in a supply chain distribution network. Therefore, it will not be discussed further. The other five modes of transportation are summarized as follows:

- *Air.* This mode of transportation is used by air carriers such as Delta Air Lines, Southwest Airlines, American Airlines, and United Airlines, which carries both passengers and cargos. Air carriers have to invest a large sum of capital in airplanes and infrastructure. Labor and fuel cost are high, and are largely trip related irrespective of the amount of cargo carried. Air carriers offer a very fast but expensive mode of transportation. They are best suited for transporting high value to weight ratio products and time-sensitive emergency shipments.

- *Highway.* This mode of transportation is used by trucking companies such as Con-way Inc., J. B. Hunt Transport Services Inc., YRC Worldwide Inc., and Arkansas Best Corporation. Trucking has relatively low fixed costs, so the barrier to entry is minimal in the industry. Trucking cost is significantly lower than that of air transport. The trucking industry can be divided into two segments: truckload (TL) and less than truckload (LTL). A TL operator generally contracts an entire trailer-load to a shipper and its pricing exhibits economies of scale. TL shipping is suited for moving a large amount of homogeneous cargo, such as the transportation between suppliers and OEMs or between manufacturing facilities and warehouses. An LTL operator collects relatively small freight from various shippers and consolidates them onto trailers. LTL shipping takes longer than TL shipping because of the need to consolidate and sort freights from different shippers. However, the cost is generally only a fraction of the cost to contract an entire trailer for exclusive shipping. It is suitable for shipments that are less than half a TL but are too large to be mailed as small packages.

- *Railroad.* This mode of transportation is used by railway companies such as BNSF Railway Company, Union Pacific Railroad, CSX Transportation Inc., and Norfolk Southern Railway. These companies have a high fixed cost in terms of rails, rail yards, locomotives, and cars. Labor and fuel cost is mainly determined by the distance travelled and the time taken rather than the number of cars. Rail cars have a very large load capacity, so the cost of rail transportation is low. However, the transportation time is long. Therefore, railroad transportation is suited for very heavy, low-value shipments that need to be carried over long distances and are not time sensitive.

- *Waterborne.* This mode of transportation is mainly used by ocean carriers such as Maersk Line, Mediterranean Shipping Company, Evergreen Line, and Neptune Orient Lines. These companies need to invest heavily to cover the cost of operations ranging from ocean liners to containers to fuel. Waterborne transportation is the dominant mode for shipping all kinds of products in global trade across the ocean. It is also used to transport large bulk of low-value shipments through inland waterways. It is the slowest of all transportation modes. However, for the large quantities shipped and the long distances travelled it is the cheapest mode of transportation. Therefore, waterborne transportation is ideally suited for shipping very large load at low cost.

- *Pipeline.* This mode of transportation is used by energy companies to transport crude oil, refined oil, and natural gas. Setting up a pipeline and the related infrastructure requires a significant upfront investment. Given this large initial investment, pipelines are only economical for regular transportation of large flow of oil and gas over long distances. Note that pneumatic pipelines using compressed air can also transport solid capsules. They are typically used to transport small valuable objects (such as cash in a bank or drugs in a hospital) in a local environment.

Companies may use a combination of different transportation modes, commonly referred to as intermodal transportation. Moving containers in a global distribution network often involves a combination of highway, water-borne, and railroad. On the other hand, small packages are shipped through package carriers that rely on intermodal transportation. Package carriers such as FedEx Corporation and United Parcel Service of America, Inc. use trucks for local delivery and pickup packages. The packages are then sorted and transported by air, full truckload on the highway, or rail to the sorting center nearest to the delivery point. From there the packages are sorted and delivered to customers using small trucks.

An emerging mode of transportation is electronic delivery of digital products. For example, Netflix, Inc. started out as a company that uses package carriers to deliver rental movies to its customers. In late 2010, it started the Internet streaming only service that allows customers to watch videos from its library that are streamed online to computers or Internet-connected televisions. Electronic delivery requires investment in information technology. It is cost effective but is only restricted to products that can be delivered in a digital format such as e-books.

When making transportation decisions, one may need to determine transportation strategies in addition to transportation modes. This is especially true when trucking is involved. One may need to decide if a *direct shipping* strategy should be used where a shipment moves directly from one location (e.g., supplier) to another (e.g., a customer or an intermediate warehouse); or a *milk run* strategy is preferred where a truck goes from one location (e.g., a warehouse) to multiple locations (e.g., retail stores). One may also use a *cross-docking* strategy where a large truck (or other high-capacity transportation modes) brings a consolidated shipment to a cross-dock point close to its final destination. The shipment is then sorted and moved to smaller trucks for local delivery. This strategy eliminates the need of intermediate warehouses while ensuring that a cheaper mode is used for long distance transportation and a faster mode is used for last mile delivery.

5.2 Frameworks for Distribution Network Design and Transportation Decision Making

In supply chain distribution network design, four major questions need to be answered: (1) What is the role of each facility in the distribution network? (2) Where should the facilities be located? (3) How much capacity should be allocated to each facility? (4) How to allocate supply sources and assign markets to each facility? To answer these questions, a company must consider a number of factors, summarized as follows:

- *Strategic factors.* A company's business strategy has a significant impact on the design of its distribution network. A company focusing on cost leadership will find low-cost locations for its manufacturing facilities even if these locations are far away from the markets they serve.

On the other hand, a company focusing on responsiveness will locate its facilities close to the markets even if the costs are high. In a global supply chain, a company may assign different strategic roles for different facilities (Ferdows 1997).

- *Technologic factors.* If the production technology requires a very large investment in building a facility (such as semiconductor manufacturing), a few high-capacity locations are desirable. On the other hand, if facilities have low fixed costs, a company should build many local facilities to reduce transportation costs.

- *Macroeconomic factors.* Macroeconomic factors include tariffs and tax incentives, exchange rates, and other economic factors that are not internal to a company. A company may choose facility locations to avoid tariffs or to take advantages of tax incentives. A company needs to build flexibility into its distribution network to help counter fluctuations in exchange rates and demand across different countries.

- *Political factors.* Companies generally prefer to build facilities in politically stable countries where ownerships and rules of commerce are well defined.

- *Infrastructure factors.* Facility locations should have good infrastructure, including labor availability, highway access, adequate utility, and proximity to transportation terminals.

- *Competitive factors.* A company must take its competitor into consideration when designing a distribution network. A fundamental decision is whether to locate the facilities close to competitors or far from them.

- *Customer service factors.* If a company's customers value a short delivery lead time, then the company needs to set up facilities close to its customers.

- *Cost factors.* The total logistics costs of a distribution network change as the number of facilities and their locations and capacities are changed. A distribution network should have a certain number of facilities that minimizes total logistics costs. The number of facilities may be increased further to shorten delivery lead time. This decision is justified if the resultant revenue increase is greater than the added facility costs.

Chopra and Meindl (2010) developed a four-phase decision framework to support distribution network design. In Phase I, a company defines its supply chain strategy in terms of the customers that it aims to satisfy. It then specifies the capability of the distribution network. By taking into account internal constraints (available capital, growth strategy, and partnership) and competitions, the company determines the overall design of the distribution network. In Phase II, the company identifies regions where facilities will be located as well as the roles and capacities of the facilities, taking into account technological factors, macroeconomic factors, political factors, and competitive factors. In Phase III, the company selects a number of potential sites within each region where its facilities are to be located, taking into account infrastructure factors. In Phase IV, the company evaluates the potential sites to identify a precise location for its facility and determines its capacity with the goal of minimizing total logistics costs.

Stank and Goldsby (2000) presented a five-level framework for transportation decision making. The framework starts with strategic, long-term decisions that focus on the overall supply chain transportation system. It then proceeds to decisions that are increasingly tactical in nature. These five levels of decisions are summarized as follows: .

- *Total network and lane design.* Long-term decisions on the transportation modes for freight movement are made, taking into account the structure of the distribution network. The decisions should specify general nature of material flows, including volume, frequency, seasonality, physical characteristics, and special handling requirements. When making these decisions, trade-offs with inventory and facility costs must be investigated.

- *Lane operation.* Lane operation decisions are made so material movements can be coordinated along inbound, facility, and outbound shipping lanes to meet service requirements at the lowest possible total costs. The primary opportunities are inbound/outbound consolidation, temporal consolidation, vehicle consolidation, and carrier consolidation. The goal is to ensure that materials arrive where they are needed in the right quantity just in time to facilitate other value-added activities while reducing transportation costs as much as possible.

- *Mode/carrier assignment.* Traditionally mode/carrier assignment decision is made using three sequential steps, namely, determination of transportation mode, identification of carrier type, and selection of individual carrier. This approach sequentially reduces the number of potential carriers until a final decision is made. Another approach is to evaluate all carriers that can meet the service criteria of a shipment and then select a carrier based on cost and availability.

- *Service negotiation.* Lowering transportation cost is a primary consideration in transportation decision making. Nonetheless, service expectations and customer- and product-specific costs must be included in contract negotiations. Legal implications of the contract should also be considered.

- *Dock-level decisions.* Dock-level decisions include load planning, routing, and scheduling. Information technology and decision-support systems play an important role in these decisions. These tools are used to make better use of transportation vehicle space, identify the most efficient routes, and schedule the utilization of equipment, facility, and personnel more cost effectively.

5.3 Distribution Network Design Models

To design a distribution network, a company needs to identify its markets and demands first. It then identifies a number of potential facility locations based on macroeconomic factors, political factors, and competitive factors. In the mean time, the capacity of each facility is determined based on technological factors. Fixed costs of keeping a facility open are then estimated. The cost of supplying one unit of product from a facility to a market is also estimated. With this information, the company can then use an optimization model to identify facility locations that minimize total costs. The input of the problem is as follows:

m: number of markets

j: index of markets, $j = 1, 2, \ldots, m$

D_j: annual demand from market j

n: number of potential facility locations

i: index of facility locations, $i = 1, 2, \ldots, n$

K_i: capacity of facility i

f_i: annual fixed cost of keeping facility i open

c_{ij}: cost of supplying one unit of product from facility i to market j

The decision variables are listed as follows:

y_i: if facility i is selected then $y_i = 1$; otherwise $y_i = 0$

x_{ij}: quantity of products in facility i used to satisfy market j

The optimization problem is formulated as follows:

$$\text{Minimize} \sum_{i=1}^{n} f_i y_i \quad \text{(facility cost)}$$

$$+ \sum_{i=1}^{n} \sum_{j=1}^{m} c_{ij} x_{ij} \quad \text{(supply cost)} \tag{5.1}$$

Subject to

$$\sum_{i=1}^{n} x_{ij} = D_j \quad \forall j \quad \text{(demand constraint)} \tag{5.2}$$

$$\sum_{j=1}^{m} x_{ij} \leq y_i K_i \quad \forall i \quad \text{(capacity constraint)} \tag{5.3}$$

$$x_{ij} \geq 0 \tag{5.4}$$

$$y_i \in \{0,1\} \tag{5.5}$$

This problem formulation is called the *capacitated facility location model*. Equation 5.2, the demand constraint, ensures that the demand from every market is satisfied. Equation 5.3, the capacity constraint, ensures that the total amount of products shipped out from a facility does not exceed its capacity. Equation 5.4 ensures that product quantities are non-negative. Equation 5.5 ensures that a facility is either not selected or selected.

Chopra and Meindl (2010) used Microsoft Excel Solver to generate a solution for the capacitated facility location model. The advantage of Excel is its friendly user interface, which allows the problem and the solution to be clearly presented. Therefore, we will use an example to illustrate how to present the capacitated facility location model in an Excel spreadsheet, using a format similar to that used by Chopra and Meindl (2010). We will then discuss how to use heuristics to develop an initial solution, followed by an attempt to use Excel Solver to improve the initial solution. Finally, we will show how Gurobi Optimizer can be reliably used to generate solution for the capacitated facility location model.

Ultra Corp. is a global laptop manufacturer. It divides its worldwide market into five regions: (1) North America, (2) South America, (3) Asia, (4) Europe, and (5) Africa. The annual demands for these five regions are 128,000, 110,000, 169,000, 111,000, and 78,000 units, respectively. It has determined that a manufacturing plant should have an annual capacity of 300,000 units. The fixed annual costs of maintaining such a manufacturing plant in the five regions are $60 million, $45 million, $50 million, $55 million, and $40 million, respectively. The costs (including production, inventory, transportation, and tariffs) of supplying one laptop from a facility location to a market are shown in Table 5.1. How can Ultra Corp. minimize its facility and supply cost?

To solve this problem, we first set up the Excel spreadsheet as shown in Figure 5.1. The costs, capacities, and demands are entered in the spreadsheet first. We then create the decision variables and set them to 0. After that

TABLE 5.1

Costs of Supplying One Laptop from a Facility Location to a Market

	Market				
Facility	North America	South America	Asia	Europe	Africa
North America	$400	$420	$450	$435	$440
South America	$380	$355	$405	$410	$400
Asia	$390	$395	$360	$375	$380
Europe	$430	$425	$415	$400	$410
Africa	$365	$370	$375	$380	$340

	A	B	C	D	E	F	G	H
1	Inputs - Costs, Capacities, Demands							
2			*Market, Supply Cost ($)*					
3	*Facility Location*	North America	South America	Asia	Europe	Africa	*Fixed Cost ($)*	*Capacity*
4	North America	400	420	450	435	440	60,000,000	300,000
5	South America	380	355	405	410	400	45,000,000	300,000
6	Asia	390	395	360	375	380	50,000,000	300,000
7	Europe	430	425	415	400	410	55,000,000	300,000
8	Africa	365	370	375	380	340	40,000,000	300,000
9	*Demand*	128,000	110,000	169,000	111,000	78,000		
10								
11	Decision Variables							
12			*Market, Production Allocation*					
13	*Facility Location*	North America	South America	Asia	Europe	Africa	*Facility Status*	
14	North America	0	0	0	0	0	0	
15	South America	0	0	0	0	0	0	
16	Asia	0	0	0	0	0	0	
17	Europe	0	0	0	0	0	0	
18	Africa	0	0	0	0	0	0	
19								
20	Constraints							
21	*Facility Location*	*Excess Capacity*						
22	North America	0						
23	South America	0						
24	Asia	0						
25	Europe	0						
26	Africa	0						
27		North America	South America	Asia	Europe	Africa		
28	*Unmet Demand*	128,000	110,000	169,000	111,000	78,000		
29								
30	Objective Function							
31	*Cost =*	$0						

Cell	Formula	Note
B22	=H4*G14-SUM(B14:F14)	Drag down to B26
F28	=B9-SUM(B14:B18)	Drag right to F28
B32	=SUMPRODUCT(G4:G8,G14:G18)+SUMPRODUCT(B4:F8, B14:F18)	

FIGURE 5.1

Spreadsheet setup for the capacitated facility location model (spreadsheet format modified from Chopra, S. and Meindl, P., *Supply Chain Management: Strategy, Planning, & Operation*, 4th edn., Pearson Prentice Hall, Upper Saddle River, NJ, 2010.)

we create the constraints by calculating the excess capacity of each facility (excess capacity must be greater than or equal to 0) and the unmet demand (unmet demand must be 0). Finally, we calculate the total cost.

Now, we use a heuristic approach to develop an initial solution. Note that the total annual demand is 596,000 units and each manufacturing plant has an annual capacity of 300,000 units. Therefore, we will need two manufacturing plants. The fixed cost for a manufacturing plant in Africa is the lowest, and the supply costs to all five regions are low. Therefore, it makes sense to build a manufacturing plant in Africa. The second and the third lowest fixed cost manufacturing plants are in South America and Asia, respectively. The annual demand in Asia is the highest and it costs the least to supply the

Asia market with an Asia manufacturing plant. Therefore, the selection of the second manufacturing plant is not that straightforward. We tentatively decide to build the second manufacturing plant in South America because of its lower facility cost.

For each manufacturing plant, it would be logical to supply regions with lower supply cost first. Therefore, the Africa plant will supply all demands from Africa (78,000 units) and North America (128,000 units). The remaining capacity of (94,000 units) is used to satisfy the demand from Asia instead of South America because it is cheaper to use the South America manufacturing plant to satisfy the demand from South America. The South America plant will supply the rest of the demand (the remaining 75,000 units from Asia, 110,000 units from South America, and 111,000 units from Europe). This solution results in total cost of $308,425,000, as shown in Figure 5.2.

We can try to use Excel Solver to see if this solution can be further improved. As shown in Figure 5.2, we tell the Solver to minimize the cost (Cell B31) by changing the decision variables (Cells B14 to G18). The decision variables must be greater than or equal to 0 ($B$14:$G$18>=0). The decision variables for facility status (selected or not selected) must be binary ($G$14:$G$18=binary). The excess capacity must be greater than or equal to 0 ($B$22:$B$26>=0). The unmet demand must be 0 ($B$28:$F$28=0).

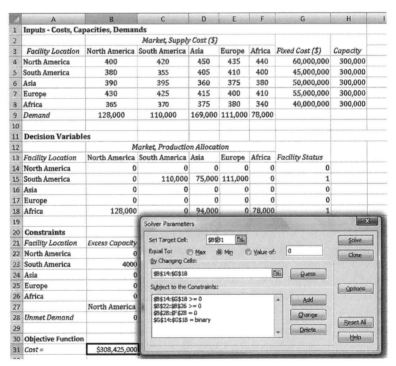

FIGURE 5.2
Excel solution for the capacitated facility location model.

The Solver was not able to further improve the initial solution. This showed that our heuristic was able to provide a good solution to the problem.

One may notice that the formulation of the capacitated facility location model looks similar to the linear programming formulation of aggregate planning problems and wonder if Gurobi Optimizer can be used to find an optimal solution. Note that Equation 5.5 requires the decision variable y_i to be either 0 or 1, which is an integer constraint. As such, the capacitated facility location model is a mixed integer linear programming model. It is an NP-hard problem (Goldreich 2010) where an optimal solution cannot be found using computationally efficient methods. Nonetheless, Gurobi Optimizer can still be used to find a very good solution quickly. The procedure for using Gurobi Optimizer to solve the capacitated facility location model is the same as that described in Chapter 3. The model file is shown as follows:

```
Minimize

\minimize the sum of facility cost and supply cost

60000000 Y1 + 45000000 Y2 + 50000000 Y3 + 55000000 Y4 + 40000000 Y5 \facility cost
+ 400 X11 + 420 X12 + 450 X13 + 430 X14 + 440 X15 \supply cost from North America
+ 380 X21 + 355 X22 + 405 X23 + 410 X24 + 400 X25 \supply cost from South America
+ 390 X31 + 395 X32 + 360 X33 + 375 X34 + 380 X35 \supply cost from Asia
+ 430 X41 + 425 X42 + 415 X43 + 400 X44 + 410 X45 \supply cost from Europe
+ 365 X51 + 370 X52 + 375 X53 + 380 X54 + 340 X55 \supply cost from Africa

Subject To

\demand constraint

X11 + X21 + X31 + X41 + X51 = 128000 \satisfy demand from North America
X12 + X22 + X32 + X42 + X52 = 110000 \satisfy demand from South America
X13 + X23 + X33 + X43 + X53 = 169000 \satisfy demand from Asia
X14 + X24 + X34 + X44 + X54 = 111000 \satisfy demand from Europe
X15 + X25 + X35 + X45 + X55 = 78000 \satisfy demand from Africa

\capacity constraint
X11 + X12 + X13 + X14 + X15 - 300000 Y1 <= 0 \constraint at North America facility
X21 + X22 + X23 + X24 + X25 - 300000 Y2 <= 0 \constraint at South America facility
X31 + X32 + X33 + X34 + X35 - 300000 Y3 <= 0 \constraint at Asia facility
X41 + X42 + X43 + X44 + X45 - 300000 Y4 <= 0 \constraint at Europe facility
X51 + X52 + X53 + X54 + X55 - 300000 Y5 <= 0 \constraint at Africa facility

Bounds

\default is >= 0

General \non-negative product quantity

X11 X12 X13 X14 X15
X21 X22 X23 X24 X25
X31 X32 X33 X34 X35
X41 X42 X43 X44 X45
X51 X52 X53 X54 X55

Binary \facility is either not selected or selected

Y1 Y2 Y3 Y4 Y5

End
```

The solution is to build one manufacturing plant in Africa and one in South America, just like our heuristic solution. However, the way the demands are satisfied is different. The Africa plant is used to fully meet the demands from Africa (78,000 units) and Europe (111,000) and partially supply the demand from Asia (111,000 units). The South America plant is used to meet the remaining demands (58,000 from Asia, 110,000 from South America, and 128,000 from North America). This solution has a total cost of $306,505,000, which is $1,920,000 lower than that of the initial solution.

After a company selects a region where its facility will be located, it needs to proceed to find a specific site for the facility. To determine the specific site, the company needs to consider its supply sources and the markets. It then estimates the quantity of components to be transported from different supply sources to the facility and the quantity of products to be transported from the facility to different markets. It also estimates the costs of transporting one unit of goods from one location to another. The locations of the facility, the supply sources, and the markets are represented as XY coordinates on a plane. Travel distances are calculated based on the coordinates. It is assumed that total transportation costs are proportional to the quantity of goods to be transported and the distance to be travelled. With this information, the company can use a *gravity location model* to find a specific site for the facility that minimizes transportation costs. The following notations are used:

n: number of supply sources and market locations

i: index of supply sources or markets, $i = 1, 2, ..., n$

(x_i, y_i): coordinates of the location of supply source or market i

f_i: cost of transporting one unit of goods for 1 mile between the facility and either supply source or market i

Q_i: quantity of goods to be transported between the facility and either supply source or market i

The decision variables are the coordinates of the facility location (x, y). The total transportation costs to be minimized are

$$TC = \sum_{i=1}^{n} f_i Q_i \sqrt{(x - x_i)^2 + (y - y_i)^2} \tag{5.6}$$

Note that this is an unconstraint optimization problem and the solution can be easily found using Microsoft Excel Solver, as demonstrated by Chopra and Meindl (2010). The procedure is demonstrated through an example as follows.

Bell Inc. plans to build a manufacturing plant in North America to produce laptops. The company purchases three main components, namely,

TABLE 5.2

Costs, Quantity, and Location Information

Supply Sources/ Markets	Transportation Costs ($/Unit/Mile)	Quantity	Coordinates	
			x_i	y_i
San Jose	0.10	296,000	150	650
Cleveland	0.15	296,000	1,700	900
Dallas	0.18	296,000	1,200	450
Los Angeles	0.35	60,108	300	500
Houston	0.35	90,527	1,250	300
New York	0.35	145,366	2,100	850

CPU, battery, and DVD drive. The CPU supplier is located at San Jose. The battery supplier is located at Cleveland. The DVD drive supplier is located at Dallas. Laptops are shipped to distribution centers at Los Angeles, Houston, and New York. The transportation costs, quantity of goods to be transported, and location coordinates are shown in Table 5.2.

We first set up the spreadsheet as shown in Figure 5.3. The costs, quantity, and location information are entered first. Next we create the decision variables and set them to 0. The distance from each supply source or market to the facility location is then calculated. The total cost is also calculated. We then

	A	B	C	D	E	F
1	Supply	Transportation	Quantity	Coordinates		
2	Sources/Markets	Costs ($/unit/mile)		x_i	y_i	*Distance*
3	San Jose	0.1	296,000	150	650	667
4	Cleveland	0.15	296,000	1,700	900	1924
5	Dallas	0.18	296,000	1,200	450	1282
6	Los Angeles	0.35	60,108	300	500	583
7	Houston	0.35	90,527	1,250	300	1285
8	New York	0.35	145,366	2,100	850	2266
9						
10	*Facility Location*					
11	x =	0				
12	y =	0				
13						
14	*Total Cost*	$341,696,165				

Cell	Formula	Note
F3	= SQRT((D3-$B$11)^2 + (E3-$B$12)^2)	Drag down to F8
B14	= SUMPRODUCT(B3:B8,C3:C8,F3:F8)	

FIGURE 5.3

Spreadsheet setup for the gravity location model (spreadsheet format modified from Chopra, S. and Meindl, P., *Supply Chain Management: Strategy, Planning, & Operation*, 4th edn., Pearson Prentice Hall, Upper Saddle River, NJ, 2010.)

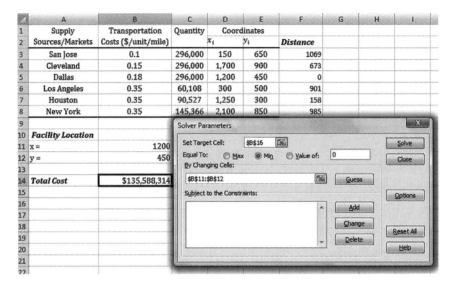

FIGURE 5.4
Excel Solver solution for the gravity location model.

invoke Excel Solver and tell the Solver to minimize the total cost (Cell B14) by changing the decision variables (Cells B11 and B12). The solution is shown in Figure 5.4. The coordinates for the optimal facility location are (1200, 450), which means Bell Inc. should build the manufacturing plant at Dallas.

5.4 Transportation Decision Making

Total network and lane design decisions require the consideration of the structure of the distribution network and inventory and facility costs. For example, a faster but higher cost transportation mode should be chosen if the result is greater savings in inventory holding cost. Lane operation decisions rely on real-time information to coordinate and consolidate shipments along various nodes in the distribution network to minimize transportation cost. For example, the same truck that is used to transport raw materials from a supplier to a manufacturing plant may also be used to transport finished product to a market if the market is located in proximity to the location of the raw material supplier. This will reduce deadhead miles that the truck needs to travel and thus reducing transportation cost. Mode/carrier assignment and service negotiations are related to supplier selection, which is usually treated as a multicriteria decision-making problem and will be discussed in Chapter 6. Dock-level decisions are micro-level decisions that are generally made with the support of specialized software tools. Here, we will discuss the details of two such decisions, namely, routing for milk run and trailer loading for direct shipping.

In a milk run, a truck departs from a central location, goes to multiple locations, and then comes back to the central location. The decision we need to make is the truck route, which is a sequence of locations that the truck visits. Two successive locations in the truck route define a path. All locations must appear in the set of paths exactly twice, once as an origination location and once as a destination location. This problem is known as the travelling salesman problem (Flood 1956). It can be formulated as an integer programming model using the following notations:

n: number of locations the truck needs to visit including the central location

i: index of the locations, $i = 1, 2, ..., n$

d_{ij}: distance between location i and location j, $i, j = 1, 2, ..., n$

The decision variables are x_{ij}. They denote the paths in the truck route. If the path from location i to location j is in the truck route, then $x_{ij} = 1$; otherwise, $x_{ij} = 0$.

The integer programming formulation is as follows:

Minimize

$$\sum_{i=1}^{n}\sum_{\substack{j=1 \\ j\neq i}}^{n} d_{ij}x_{ij} \tag{5.7}$$

Subject to

$$\sum_{\substack{i=1 \\ i\neq j}}^{n} x_{ij} = 1 \quad \forall j \tag{5.8}$$

$$\sum_{\substack{j=1 \\ j\neq i}}^{n} x_{ij} = 1 \quad \forall i \tag{5.9}$$

$$x_{ij} + x_{ji} \leq 1, \quad 1 \leq i \leq n-1, \quad j > i \tag{5.10}$$

$$x_{ij} \in \{0,1\} \tag{5.11}$$

Equation 5.7 minimizes the total travel distance. Equation 5.8 ensures that for each destination location in a path there is one and only one origination location. Equation 5.9 ensures that for each origination location in a path there is one and only one destination location. Equation 5.10 ensures that two identical paths (a path from location i to location j is identical to a path

TABLE 5.3

Distance Matrix for Farmer's Dairy Milk Run

	Farmer's Dairy (1)	EZ-Mart (2)	Kwik-E-Mart (3)	Speedy Mart (4)	Hearty Mart (5)	Green Mart (6)
Farmer's Dairy (1)	—	5.3	7.4	7.0	9.7	17.1
EZ-Mart (2)	5.3	—	9.3	12.1	14.9	18.4
Kwik-E-Mart (3)	7.4	9.3	—	7.6	14.4	24.7
Speedy Mart (4)	7.0	12.1	7.6	—	7.8	23.8
Hearty Mart (5)	9.7	14.9	14.4	7.8	—	15.6
Green Mart (6)	17.1	18.4	24.7	23.8	15.6	—

from location j to location i) cannot appear in the truck route more than once. Equation 5.11 ensures that a path is either in the truck route or not in the truck route.

Through the following example, we will show how to solve this routing problem using Gurobi Optimizer and how to interpret the result. A dairy producer, Farmer's Dairy, supplies dairy products to five local convenience stores, EZ-Mart, Kwik-E-Mart, Speedy Mart, Hearty Mart, and Green Mart. The distances among these locations are shown in Table 5.3. Delivery is made each day using a single truck, which has sufficient capacity to carry all the required products. To determine the optimal truck route, we denote Farmer's Dairy, EZ-Mart, Kwik-E-Mart, Speedy Mart, Hearty Mart, and Green Mart as locations 1, 2, 3, 4, 5, and 6, respectively, and develop the following Gurobi model:

```
Minimize

\minimize total travel distance
                5.3 x12 +   7.4 x13 +   7.0 x14 +   9.7 x15 + 17.1 x16
+   5.3 x21 +               9.3 x23 + 12.1 x24 + 14.9 x25 + 18.4 x26
+   7.4 x31 +   9.3 x32 +               7.6 x34 + 14.4 x35 + 24.7 x36
+   7.0 x41 + 12.1 x42 +   7.6 x43 +               7.8 x45 + 23.8 x46
+   9.7 x51 + 14.9 x52 + 14.4 x53 +   7.8 x54 +             15.6 x56
+ 17.1 x61 + 18.4 x62 + 24.7 x63 + 23.8 x64 + 15.6 x65


Subject To

\one and only one origination location in a path (a total of six paths)
        x21 + x31 + x41 + x51 + x61 = 1
x12 +         x32 + x42 + x52 + x62 = 1
x13 + x23 +         x43 + x53 + x63 = 1
x14 + x24 + x34 +         x54 + x64 = 1
x15 + x25 + x35 + x45 +         x65 = 1
x16 + x26 + x36 + x46 + x56         = 1
```

(continued)

```
\one and only one destination location in a path (a total of six paths)
        x12 + x13 + x14 + x15 + x16 = 1
x21 +        + x23 + x24 + x25 + x26 = 1
x31 + x32 +        x34 + x35 + x36 = 1
x41 + x42 + x43 +        x45 + x46 = 1
x51 + x52 + x53 + x54 +       x56 = 1
x61 + x62 + x63 + x64 + x65       = 1

\identical paths cannot appear in the route more than once
x12 + x21 <= 1
x13 + x31 <= 1
x14 + x41 <= 1
x15 + x51 <= 1
x16 + x61 <= 1
x23 + x32 <= 1
x24 + x42 <= 1
x25 + x52 <= 1
x26 + x62 <= 1
x34 + x43 <= 1
x35 + x53 <= 1
x36 + x63 <= 1
x45 + x54 <= 1
x46 + x64 <= 1
x56 + x65 <= 1

Bounds

\default is >= 0

Binary

    x12 x13 x14 x15 x16
x21     x23 x24 x25 x26
x31 x32     x34 x35 x36
x41 x42 x43     x45 x46
x51 x52 x53 x54     x56
x61 x62 x63 x64 x65

End
```

The solution found by Gurobi Optimizer is $x_{13} = 1$, $x_{21} = 1$, $x_{34} = 1$, $x_{45} = 1$, $x_{56} = 1$, and $x_{62} = 1$, with a total travel distance of 62.1 miles. As shown in Figure 5.5, the solution indicates that the truck route consists of the following six paths: (1) from Farmer's Dairy to Kwik-E-Mart, (2) from EZ-Mart to Farmer's Dairy, (3) from Kwik-E-Mart to Speedy Mart, (4) from Speedy Mart to Hearty Mart, (5) from Hearty Mart to Green Mart, and (6) from Green Mart to EZ-Mart. We can see that these six paths form a single loop that covers all the locations, which is a valid truck route for the milk run from Farmer's Dairy.

In direct shipping, a truck travels directly from one location to another so the travel distance is fixed. Given the amount of freights that need to be transported, the goal is to minimize the number of trucks used. This is equivalent to minimizing the wasted floor space of a truck trailer, which is commonly referred to as the trailer loading problem. Here, we consider a situation

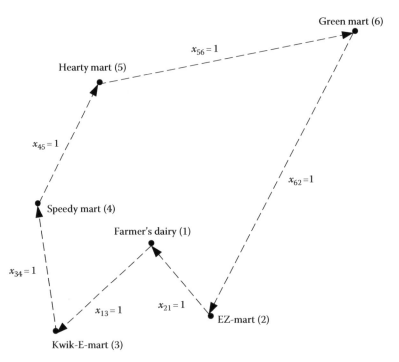

FIGURE 5.5
Graphical illustration of the solution for Farmer's Dairy milk run.

where freights from multiple suppliers need to be loaded to a trailer. Supplier freights are rectangular boxes, each having the following properties:

- Box type: rack or skid
- Dimension: length, width, and height
- Stackability: other boxes can or cannot be stacked on top of the box
- Rotatability: both length and width can be aligned with the back of the trailer or only length can be aligned with the back of the trailer

When loading supplier freights, the following constraints must be satisfied:

- One supplier's boxes cannot be stacked on another supplier's boxes
- Boxes can only rotate on the base (rotate length and width)
- Racks can stack only on racks that have the same base size
- Skids can stack only on skids with equal or larger base size

It is common to arrange the boxes into a set of towers so the trailer loading problem can be simplified as a two-dimensional packing problem (Lodi et al. 2002). A trailer and its cargo (loaded towers) can then be represented

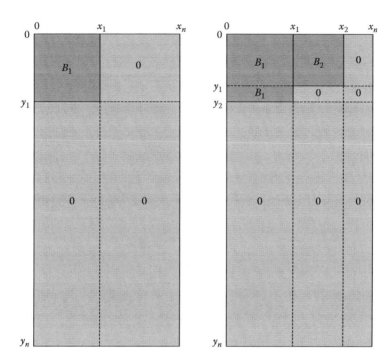

FIGURE 5.6
Matrix representation of a trailer and its cargo.

as a matrix using the following procedure. The empty trailer is represented using a 1×1 matrix $\mathbf{M}_0$, where $m_{11} = 0$, indicating empty space. As shown in Figure 5.6, when a box B_1 is placed inside the upper-left corner of the trailer, $\mathbf{M}_0$ expands to a new 2×2 matrix $\mathbf{M}_1$, where $m_{11} = B_1$ (indicating it is occupied by box B_1) and $m_{12} = 0$, $m_{21} = 0$, $m_{22} = 0$ (indicating that they are empty spaces). When another box B_2 is placed next to B_1 at the space represented by m_{12}, $\mathbf{M}_1$ expands to a new 3×3 matrix $\mathbf{M}_2$, where $m_{11} = m_{21} = B_1$ (indicating both spaces are occupied by box B_1), $m_{12} = B_2$ (indicating it is occupied by box B_2), and the rest of the matrix elements are 0 (indicating they are empty spaces). Note that when $y_1 = y_2$, $\mathbf{M}_2$ will be a 3×2 matrix instead of a 3×3 matrix. Matrix $\mathbf{M}_i$ is accompanied by two vectors $\mathbf{X}_i$ and $\mathbf{Y}_i$ to record the position of each element of the matrix. For the example shown in Figure 5.6, $\mathbf{X}_0 = [x_n]$, $\mathbf{Y}_0 = [y_n]$; $\mathbf{X}_1 = [x_1, x_n]$, $\mathbf{Y}_1 = [y_1, y_n]$; $\mathbf{X}_2 = [x_1, x_2, x_n]$, $\mathbf{Y}_2 = [y_1, y_2, y_n]$. In this way, $\mathbf{M}_i$, $\mathbf{X}_i$, and $\mathbf{Y}_i$ uniquely define a trailer with its cargos.

A heuristic algorithm for trailer loading is shown in Figure 5.7. After tower building, the algorithm uses a one-step look ahead heuristic to reduce wasted space. The idea is to try putting two smaller towers (in terms of base area) instead of a larger one in an empty space, if putting the larger tower leaves no room for another tower. Rotation of the tower base, if permissible, is attempted while doing so. The components of the algorithm are discussed in detail as follows.

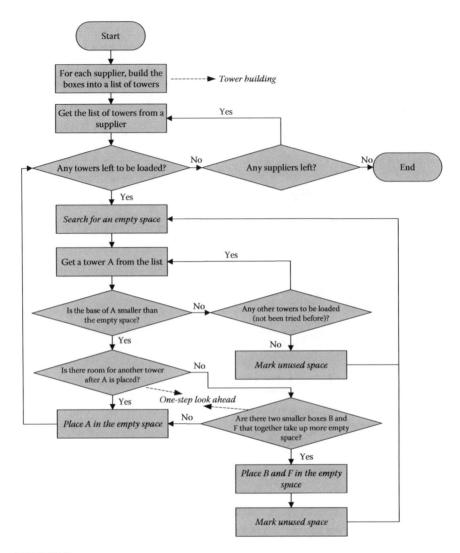

FIGURE 5.7
Trailer loading heuristic algorithm flowchart.

Tower building: As mentioned previously, there are two types of boxes, namely, skids and racks. A skid with a smaller base can go on top of one with a larger base, whereas a rack can only go on top of one with the same base. This is the only difference in building towers for skids and racks. Therefore, we will discuss the procedure of tower building for skids. The extension to tower building for racks is straightforward. Let $S = \{s_1, s_2, ..., s_{n_S}\}$ be the skids from a supplier arranged in descending order of their base area, where n_S is the number of skids. Let L_i, W_i, and H_i denote the length, width, and height of skid s_i, respectively. Let K_i denote the stackability of skid s_i. If other skids can sit on top of skid s_i, then K_i = TRUE; otherwise, K_i = FALSE. Let H denote the height of the trailer. The pseudo codes for building a tower t for a set of skids are as follows:

```
//find a skid as the base of the tower, the stackability of this skid must
//be TRUE
i = 1
while (Kᵢ = FALSE and i < nₛ) {
      i = i + 1
}
if i = nₛ then each skid forms a tower, exit    //these skids are not stackable
else    //use the skid as the tower base
      t = ∅
      α = i, t = t ∪ {i}
      S = {sₐ₊₁, ..., sₙₛ}//consider only skids with smaller base area
      h = H    //allowable height is the trailer height
      while (α ≠ ∅ and S ≠ ∅) {
            h = h − Hₐ    //subtract the height of the current skid in the
                          //tower set from the allowable height
            τ = α
            α = FindFit(α, h, S)    //find the next skid that can sit on top of
                                    //the current skid
            if α ≠ ∅ then
                  S = {sₐ₊₁, ..., sₙₛ}
                  if Kₐ = TRUE then
                        t = t ∪ {α}
                  else    //if the skid is non-stackable
                        β = FindFitStack(τ, h, S)    //find a stackable skid instead
                        if β ≠ ∅ then
                              S = {sᵦ₊₁, ..., sₙₛ}
                              γ = FindFit(β, h−Hᵦ, S)    //find the next skid that
                                                        //can sit on top
                              if γ = ∅ then    //if not found, the non-stackable
                                               //skid is the top of the tower
                                    t = t ∪ {α}
                                    α = ∅
                              else    //otherwise, use the stackable skid as the
                                      //next layer of the tower
                                    α = β
                                    t = t ∪ {α}
                        else    //no stackable skid found, the non-stackable is
                                //the top of the tower
                              t = t ∪ {α}
                              α = ∅
      }
```

The pseudo codes for the function that finds a skid from the set **S** that can sit on top of skid B considering the trailer height h, FindFit(B, h, **S**), are as follows:

```
p = Ø
if S = Ø return p
for i = 1 to n_S
        if L_i < L_B and W_i < W_B then    //for rack the condition would be L_i =
                                           //L_B and W_i = W_B
            if H_i < h
                    return i
    return p
```

The pseudo codes for the function that finds a stackable skid from the set **S** that can sit on top of skid B considering the trailer height h, FindFitStack(B, h, **S**), are as follows:

```
p = Ø
if S = Ø return p
for i = 1 to n_S
        if L_i < L_B and W_i < W_B then    //for rack the condition would be L_i =
                                           //L_B and W_i = W_B
            if H_i < h and K_i = TRUE
                    return i
    return p
```

In addition to its elements, each tower set **t** is characterized by the following:

- Length and width, which are defined as the length and width, respectively, of its base box
- Rotatability, which is the rotatability of the base box

Note that the aforementioned routine builds one tower at a time (unless all the skids considered are nonstackable). After one tower **t** is built, the elements of **t** are removed from the skid set **S**. The routine repeats until the skid set **S** is empty. Towers are built for skids first and then for racks. The resulting towers are sequentially arranged in a list (by putting tower built first at the beginning of the list, followed by the next tower built, and so forth).

Search for an empty space: An empty space is a rectangle defined by the coordinate of its upper-left corner (x_s, y_s), its length along the X-axis, X_e, and its length along the Y-axis, Y_e. Note that in matrix **M**, an empty space may be represented by several adjacent elements with a value of 0. The empty trailer is represented by $\mathbf{M} = [0]$, $\mathbf{X} = \{W\}$, and $\mathbf{Y} = \{L\}$, in which W and L are the width and length of the trailer, respectively. The pseudo codes for finding an empty space are as follows:

Sequentially (left to right, top to bottom) search for the first element in matrix $\mathbf{M}$ ($n_x \times n_y$) whose value is 0, denoted as $m_{pq} = 0$
$P_s = p$, $Q_s = q$ //to be used for marking unused space and placing tower
                //in empty space
if $p > 1$ then $x_s = x_{p-1}$ else $x_s = 0$
if $q > 1$ then $y_s = y_{q-1}$ else $y_s = 0$
$Y_e = y_{n_y} - y_s$ //the length along the Y axis always extends to the end
                //of the trailer
while ($p < n_x$ and $m_{pq} = 0$) {
        $p = p + 1$

}
$X_e = x_p - x_s$ //the length along the X axis extends to a cell that is
                //occupied
$R_s = p$ //to be used for marking unused space

Mark unused space: An unused space is a space that is too small to load any boxes. It should be marked (as –1) so the algorithm does not attempt to use it. The pseudo codes for marking unused space are as follows:

$q_l = Q_s$, $q_r = Q_s$
//find the end of an obstacle at Y axis from the left
if $P_s > 1$ then
        while ($q_l \leq n_y$ and $m_{(P_s-1)q_l} \neq 0$) {
                $q_l = q_l + 1$
        }
else
        $q_l = n_y + 1$
//find the end of an obstacle at Y axis from the right
if $R_s < n_x$ then
        while ($q_r \leq n_y$ and $m_{(R_s+1)q_r} \neq 0$) {
                $q_r = q_r + 1$
        }
else
        $q_r = n_y + 1$
$q = \min\{q_l, q_r\}$
$m_{uv} = -1$, where $P_s \leq u \leq R_s$, $Q_s \leq v < q$

One-step look ahead: Let $\mathbf{T} = \{1, 2, \ldots, n_T\}$ denote the set of remaining towers after A is selected (A is always selected from the beginning of the tower list).

The one-step look ahead heuristic is invoked if the number of elements in **T** is greater than 1. The pseudo codes are as follows:

f_A = BaseSmaller(A, X_e, Y_e)   //X_e and Y_e are obtained from the search for
                                         //empty space routine
if f_A > 0 then   //start one-step look ahead heuristic
    $g = L_A$
    if f_A = 2 then $g = W_A$
    $x_l = X_e - g$ //calculate remaining empty space if A is placed
    $U = T - \{A\}$
    f_C = FindTower(U, x_l, Y_e, F)   //find the next tower that fits the
                                             //remaining empty space
    if f_C > 0 then place A in the empty space (rotate A if f_A = 2)
    else
            let D be the element in **U** that has the smallest length considering
                rotatability, let $U = U - \{D\}$
            f_B = FindTower(U, $X_e - L_D$, Y_e, B)   //find a tower B that together
                                                          //with D fits the empty space
            if f_B > 0 then   //find the largest tower that together with B fits
                                 //the empty space
                $x_e = X_e - L_B$
                if f_B = 2 then $x_e = X_e - W_B$
                $U = T - \{A, B\}$
                f_F = FindTower(U, x_e, Y_e, F)
                if f_F = 1 then $x_e = x_e - L_F$
                else $x_e = x_e - W_F$
                if $x_e < x_l$ then   //B and F occupy more space than A
                        place B and F in the empty space (rotate B if f_B = 2,
                            rotate F if f_F = 2)
                else
                        place A in the empty space (rotate A if f_A = 2)
            else
                place A in the empty space (rotate A if f_A = 2)

The pseudo codes for finding if a tower A can fit into an empty space (X_e, Y_e), BaseSmaller(A, X_e, Y_e), are as follows:

if $L_A \leq X_e$ and $W_A \leq Y_e$ then
    return 1
else
    if r_A = TRUE and $W_A \leq X_e$ and $L_A \leq Y_e$ then   //rotate the tower if necessary
        return 2
    else
        return 0

The pseudo codes for finding the first tower F from a tower set **U** that can fit into an empty space (X_e, Y_e), FindTower(**U**, X_e, Y_e, F), are as follows:

```
{
    let F be the first element of U, let U = U − {F}    //U is passed as a
                                                        //local copy

    f = BaseSmaller(F, X_e, Y_e)
} while (U ≠ φ and f = 0)
return f
```

Place a tower in the empty space: Let L_B and W_B denote the length and width of tower B, respectively. It is assumed that the length of the tower is always aligned with the width (X-axis) of the trailer (the loading algorithm has already attempted rotation of the tower so this assumption does not compromise the optimality of the final loading solution). The pseudo codes for loading the tower B to a trailer represented by **M**, **X**, and **Y** and placing it in the empty space represented starting at cell (P_s, Q_s) are as follows:

```
//find the coordinate of the upper left corner of the empty space
if P_s > 1 then O_x = x_{P_s−1} else O_x = 0
if Q_s > 1 then O_y = y_{Q_s−1} else O_y = 0
k = P_s, l = Q_s
if O_x + L_B ≠ x_{P_s} then    //insert a column
        find the order of O_x + L_B in X, assign it to k
        insert the value O_x + L_B to X at the kth position
        make a copy of the kth row of M and insert it after the kth row
if O_y + W_B ≠ y_{Q_s} then    //insert a row
        find the order of O_y + W_B in Y, assign it to l
        insert the value O_y + W_B to Y at the lth position
        make a copy of the lth column of M and insert it after the lth column
m_{uv} = B, where P_s ≤ u ≤ k, Q_s ≤ v ≤ l
```

5.5 Case Studies

Case Study 5.5.1

NJ Closets is a New Jersey–based company that manufactures closet organizers. It has invested heavily in market development over the past 2 years. As a result, the demand for its product from the West Coast has increased significantly. The company estimated that the demand will grow to an average of 15,000 closet organizers per week quickly before leveling off. It was also estimated that weekly demand is independent and follows a normal distribution with a standard deviation of 3000 closet organizers. The company is now planning to establish a warehouse to supply the West Coast. Two locations are under consideration, one in San Francisco and the other in Los Angeles. Closet organizers will be shipped from the warehouse to customers by a LTL trucking company. The cost is estimated as $.45 and $.60 per closet organizer from San Francisco and Los Angeles, respectively. The annual cost for operating a warehouse is $500,000 + 2x in San Francisco and $400,000 + 2x in Los Angeles, where x is the storage capacity of the warehouse. Closet organizers can be transported from New Jersey to the warehouse either by railroad with a lead time of 3 weeks or by highway with a lead time of 1 week. The cost for railroad transportation is estimated as $.40 per closet organizer, whereas the cost for highway transportation is estimated as $1 per closet organizer. NJ Closet has determined that it will supply the warehouse every 4 weeks. The inventory holding cost is $50 per closet organizer per year and the desired *CSL* is 95%. Should the company use railroad transportation or highway transportation? Should the company establish the warehouse in San Francisco or in Los Angeles?

NJ Closets needs to minimize the sum of warehouse operating cost, transportation cost, and inventory holding cost. The cost for railroad transportation is lower than that for highway transportation. However, when railroad transportation is used, a higher level of safety inventory is required because the replenishment lead time is longer than that with highway transportation. This would increase inventory holding cost and warehouse operating cost because a warehouse with larger storage capacity is required. Because the inbound transportation (from New Jersey manufacturing plant to the warehouse) cost and lead time are the same for both San Francisco and Los Angeles, the decision of using railroad transportation or highway transportation can be made independent of the warehouse location.

NJ Closets plans to supply the warehouse every 4 weeks. This implies that the company uses a periodic review policy. The safety inventory required when using highway transportation is $F_S^{-1}(0.95) \times 3{,}000 \times \sqrt{4+1} \approx 11{,}034$. On the other hand, the safety inventory required when using railroad transportation is $F_S^{-1}(0.95) \times 3{,}000 \times \sqrt{4+3} = 13{,}056$. The difference is $13{,}056 - 11{,}034 = 2{,}022$ closet organizers. The increase in inventory holding cost is $2{,}022 \times \$50 = \$101{,}100$ per year, whereas the increase in warehouse operating cost is $2{,}022 \times \$2 = \$4{,}044$ per year. However, assuming there are 52 weeks in a year, the saving from railroad transportation is $52 \times 15{,}000 \times (1 - 0.4) = \$468{,}000$. This is higher than the increase in inventory holding and warehouse operating cost. Therefore, using railroad transportation is a better choice. It allows NJ Closets to achieve a yearly saving of $468{,}000 - 101{,}100 - 4{,}044 = \$362{,}856$.

Now we consider the location of the warehouse. Because the inbound (from NJ Closets' manufacturing plant to the warehouse) transportation costs are the same for both locations, we only need to consider outbound transportation (from the warehouse to customers) cost and warehouse operating cost. Because the inbound transportation lead time is the same for both locations, the required safety inventory is the same, which means the warehouse capacity is the same for both locations. Therefore, the annual warehouse operating cost is $100,000 higher in San Francisco than that in Los Angeles. However, when the warehouse is located in San Francisco, the annual outbound transportation cost is $52 \times 15,000 \times (0.60 - 0.45) = \$117,000$ lower than when the warehouse is located in Los Angeles. Therefore, NJ Closets should operate a warehouse in San Francisco.

Case Study 5.5.2

MD Techtronic is a medical equipment manufacturer that has two manufacturing plants, one in Austin, Texas, and the other in Charlotte, North Carolina. The company has recently acquired a maker of portable x-ray machines and would like to move the production to its own manufacturing plants. Both the Austin plant and the Charlotte plant can produce up to 30,000 x-ray machines a year. Because MD Techtronic produces a variety of other products in these plants, no fixed annual cost (overhead) is separately attributed to the production of the x-ray machines. The x-ray machines are sold worldwide in five major markets: Asia-Pacific with an annual demand of 15,000 units, Europe with an annual demand of 10,000 units, North America with an annual demand of 8,000 units, Middle East with an annual demand of 4,000 units, and South America with an annual demand of 3,000 units. The x-ray machines sold in different markets differ in terms of their power outlet as well as the language of the instruction manuals. These market customization works will be done in separate distribution centers. There are three potential locations for the distribution centers, namely, Houston, New York, and Los Angeles. Each of these distribution centers has an annual processing capacity of 40,000 machines. The annual cost for operating a distribution center is $1.5 million in Houston, $2 million in New York, and $1.8 million in Los Angeles. The cost for transporting one x-ray machine from a manufacturing plant to a distribution center and that from a distribution center to a market is shown in Table 5.4. A key raw material for producing the x-ray machines is Thulium, a rare earth element. There are three qualified suppliers for Thulium, located in China, Australia, and Kazakhstan, respectively. The cost of supplying Thulium needed for producing one x-ray machine from a supplier to a manufacturing plant is shown in Table 5.5. All of the suppliers have sufficient capacity to meet MD Techtronic's needs. How should MD Techtronic allocate its production of x-ray machines? Where should the distribution center(s) be located? What supplier(s) should the company use?

In this case, we need to simultaneously consider suppliers, manufacturing plants, distribution centers, markets, and the quantity of goods shipped from one location to another. We can formulate this problem as a mixed integer linear programming model. Because no separate fixed annual cost is attributed to the production of x-ray machines in a

TABLE 5.4

Transportation Cost to and from Distribution
Centers

	Houston	New York	Los Angeles
Austin	$20	$100	$70
Charlotte	$90	$30	$120
Asia-Pacific	$220	$250	$200
Europe	$195	$180	$210
North America	$70	$75	$80
Middle East	$240	$230	$235
South America	$90	$115	$95

TABLE 5.5

Raw Material Supply Cost

	China	Australia	Kazakhstan
Austin	$150	$175	$165
Charlotte	$160	$170	$180

manufacturing plant, the decision variables that we need to consider are
as follows:

yH, yN, yL: whether a distribution center should be located in
Houston, New York, or Los Angeles, respectively

xIA, xUA, xKA: quantity of Thulium supplied to the Austin plant
from China, Australia, and Kazakhstan, respectively

xIC, xUC, xKC: quantity of Thulium supplied to the Charlotte
plant from China, Australia, and Kazakhstan, respectively

xAH, xAN, xAL: quantity of x-ray machines shipped from the
Austin plant to the distribution center in Houston, New York,
or Los Angeles, respectively

xCH, xCN, xCL: quantity of x-ray machines shipped from the
Charlotte plant to the distribution center in Houston, New
York, or Los Angeles, respectively

xHP, xHE, xHO, xHM, xHS: quantity of x-ray machines shipped
from the Houston distribution center to Asia-Pacific, Europe,
North America, Middle East, and South America, respectively

xNP, xNE, xNO, xNM, xNS: quantity of x-ray machines shipped
from the New York distribution center to Asia-Pacific, Europe,
North America, Middle East, and South America, respectively

xLP, xLE, xLO, xLM, xLS: quantity of x-ray machines shipped from
the Los Angeles distribution center to Asia-Pacific, Europe,
North America, Middle East, and South America, respectively

The objective is to minimize the sum of distribution center cost, supply cost, and transportation cost (from the manufacturing plants to the

distribution centers and from the distribution centers to the markets). The constraints to be satisfied are as follows:

- *Demand constraint.* Demands from the five markets must be satisfied from the distribution centers.
- *Capacity constraint.* The number of x-ray machines transported from a distribution center cannot exceed the capacity of the distribution center. Similarly, the number of x-ray machines transported from a manufacturing plant cannot exceed the capacity of the manufacturing plant.
- *Material constraint.* The Thulium supplied to a manufacturing plant must be more than or equal to the amount needed for production (i.e., the number of x-ray machines to be transported from the manufacturing plant). Similarly, the number of x-ray machines transported to a distribution center must be more than or equal to the amount transported out from the distribution center.

The Gurobi model for solving this problem is shown as follows:

```
Minimize

\minimize the sum of facility cost, supply cost, and shipping cost
1500000 yH + 2000000 yN + 1800000 yL \fixed cost of distribution centers
+ 150 xIA + 175 xUA + 165 xKA \supply cost to Austin plant
+ 160 xIC + 170 xUC + 180 xKC \supply cost to Charlotte plant
+ 20 xAH + 100 xAN + 70 xAL \transportation from Austin plant
+ 90 xCH + 30 xCN + 120 xCL \transportation from Charlotte plant
+ 220 xHP + 195 xHE + 70 xHO + 240 xHM + 90 xHS \transportation from Houston
+ 250 xNP + 180 xNE + 75 xNO + 230 xNM + 115 xNS \transportation from New York
+ 200 xLP + 210 xLE + 80 xLO + 235 xLM + 95 xLS \transportation from Los Angeles

Subject To

\demand constraint
xHP + xNP + xLP = 15000 \from Asia-Pacific
xHE + xNE + xLE = 10000 \from Europe
xHO + xNO + xLO = 8000 \from North America
xHM + xNM + xLM = 4000 \from Middle East
xHS + xNS + xLS = 3000 \from South America

\capacity constraint
xHP + xHE + xHO + xHM + xHS - 40000 yH <= 0 \at Houston DC
xNP + xNE + xNO + xNM + xNS - 40000 yN <= 0 \at New York DC
xLP + xLE + xLO + xLM + xLS - 40000 yL <= 0 \at Los Angeles DC
xAH + xAN + xAL <= 30000 \at Austin plant
xCH + xCN + xCL <= 30000 \at Charlotte plant

\material constraint: quantity in must be more than or equal to quantity out
xIA + xUA + xKA - xAH - xAN - xAL >= 0 \raw material for Austin plant
xIC + xUC + xKC - xCH - xCN - xCL >= 0 \raw material for Charlotte plant
xAH + xCH - xHP - xHE - xHO - xHM - xHS >= 0 \product for Houston DC
xAN + xCN - xNP - xNE - xNO - xNM - xNS >= 0 \product for New York DC
xAL + xCL - xLP - xLE - xLO - xLM - xLS >= 0 \product for Los Angeles DC

Bounds

\default is >= 0

General \non-negative raw material and product quantity

xIA xUA xKA xIC xUC xKC
xAH xAN xAL xCH xCN xCL
```

(continued)

(continued)

```
xHP xHE xHO xHM xHS
xNP xNE xNO xNM xNS
xLP xLE xLO xLM xLS

Binary \distribution center is either not selected or selected

yH yN yL

End
```

The solution found by Gurobi Optimizer is as follows. MD Techtronic should produce 30,000 x-ray machines in the Austin manufacturing plant and 10,000 x-ray machines in the Charlotte manufacturing plant. Only one distribution center should be established, at Houston. All Thulium needed should be supplied from China. The total cost for this solution is $16,140,000.

Case Study 5.5.3

Speedy Gourmet manufactures three lines of frozen dinner, namely, Shanghai Mama's Asian Cuisine, En Casa Mexican Cuisine, and Di Pescara Italian Gourmet. The company has two manufacturing plants, one in Dayton, Ohio, and the other in Provo, Utah; each has an annual production capacity of 2.6 million pounds of frozen dinner. The unit production costs for all three lines of frozen dinner are the same in the two manufacturing plants. However, each line of frozen dinner at a plant is managed by a separate department and incurs a fixed cost (overhead) of $100,000 per year. For example, if the Provo plant produces all three lines, there will be an annual overhead of $300,000 in addition to production cost. On the other hand, if only one line of frozen dinner is produced, then the annual overhead is only $100,000. Speedy Gourmet also has four distribution centers, located in Philadelphia, Chicago, New Orleans, and Las Vegas. The annual demand of the three lines of frozen dinner from these distribution centers is shown in Table 5.6. The cost of transporting 1000 lb of frozen dinner from Dayton is $135 to Philadelphia, $75 to Chicago, $215 to New Orleans, and $488 to Las Vegas. The cost of transporting 1000 lb of frozen dinner from Provo is $540 to Philadelphia, $355 to Chicago, $428 to New Orleans, and $95 to Las Vegas. How should Speedy Gourmet plan its production to supply the four distribution centers?

TABLE 5.6

Annual Demand of Frozen Dinners (in Thousand lb)

	Dinner Line		
Distribution Center	**Asian Cuisine**	**Mexican Cuisine**	**Italian Gourmet**
Philadelphia	500	180	320
Chicago	350	170	280
New Orleans	150	300	250
Las Vegas	400	450	200

Because the unit production costs for all three lines of frozen dinner are the same in the two manufacturing plants, Speedy Gourmet needs to minimize the sum of overhead cost and transportation cost. The decision variables are as follows:

yDA, yDM, yDI: whether the Dayton manufacturing plant should produce Asian Cuisine, Mexican Cuisine, and Italian Gourmet lines of frozen dinner, respectively

yPA, yPM, yPI: whether the Provo manufacturing plant should produce Asian Cuisine, Mexican Cuisine, and Italian Gourmet lines of frozen dinner, respectively

xADH, xADC, xADN, xADL: the quantity of Asian Cuisine to be supplied from the Dayton manufacturing plant to Philadelphia, Chicago, New Orleans, and Las Vegas distribution center, respectively

xMDH, xMDC, xMDN, xMDL: the quantity of Mexican Cuisine to be supplied from the Dayton manufacturing plant to Philadelphia, Chicago, New Orleans, and Las Vegas distribution center, respectively

xIDH, xIDC, xIDN, xIDL: the quantity of Italian Cuisine to be supplied from the Dayton manufacturing plant to Philadelphia, Chicago, New Orleans, and Las Vegas distribution center, respectively

xAPH, xAPC, xAPN, xAPL: the quantity of Asian Cuisine to be supplied from the Provo manufacturing plant to Philadelphia, Chicago, New Orleans, and Las Vegas distribution center, respectively

xMPH, xMPC, xMPN, xMPL: the quantity of Mexican Cuisine to be supplied from the Provo manufacturing plant to Philadelphia, Chicago, New Orleans, and Las Vegas distribution center, respectively

xIPH, xIPC, xIPN, xIPL: the quantity of Italian Cuisine to be supplied from the Provo manufacturing plant to Philadelphia, Chicago, New Orleans, and Las Vegas distribution center, respectively

The demand from the four distribution centers must be satisfied from the two manufacturing plants. The quantity of frozen dinners to be produced in a manufacturing plant cannot exceed the plant's production capacity. The Gurobi model for solving this problem is shown as follows:

```
Minimize

\minimize the sum of overhead cost and transportation cost
    100000 yDA + 100000 yDM + 100000 yDI \overhead at Dayton
+ 100000 yPA + 100000 yPM + 100000 yPI \overhead at Provo
+ 135 xADH + 135 xMDH + 135 xIDH \transportation cost from Dayton to Philadelphia
+ 75  xADC + 75  xMDC + 75  xIDC \transportation cost from Dayton to Chicago
+ 215 xADN + 215 xMDN + 215 xIDN \transportation cost from Dayton to New Orleans
+ 488 xADL + 488 xMDL + 488 xIDL \transportation cost from Dayton to Las Vegas
+ 540 xAPH + 540 xMPH + 540 xIPH \transportation cost from Provo to Philadelphia
```

(continued)

(continued)

```
 + 355 xAPC + 355 xMPC + 355 xIPC \transportation cost from Provo to Chicago
 + 428 xAPN + 428 xMPN + 428 xIPN \transportation cost from Provo to New Orleans
 + 95  xAPL + 95  xMPL + 95  xIPL \transportation cost from Provo to Las Vegas

Subject To

\demand constraint
xADH + xAPH = 500 \Asian Cuisine to Philadelphia
xADC + xAPC = 350 \Asian Cuisine to Chicago
xADN + xAPN = 150 \Asian Cuisine to New Orleans
xADL + xAPL = 400 \Asian Cuisine to Las Vegas
xMDH + xMPH = 180 \Mexican Cuisine to Philadelphia
xMDC + xMPC = 170 \Mexican Cuisine to Chicago
xMDN + xMPN = 300 \Mexican Cuisine to New Orleans
xMDL + xMPL = 450 \Mexican Cuisine to Las Vegas
xIDH + xIPH = 320 \Italian Gourmet to Philadelphia
xIDC + xIPC = 280 \Italian Gourmet to Chicago
xIDN + xIPN = 250 \Italian Gourmet to New Orleans
xIDL + xIPL = 200 \Italian Gourmet to Las Vegas

\capacity constraint
xADH + xADC + xADN + xADL - 2600 yDA <= 0 \Dayton Asian Cuisine
xMDH + xMDC + xMDN + xMDL - 2600 yDM <= 0 \Dayton Mexican Cuisine
xIDH + xIDC + xIDN + xIDL - 2600 yDI <= 0 \Dayton Italian Gourmet
xAPH + xAPC + xAPN + xAPL - 2600 yPA <= 0 \Provo Asian Cuisine
xMPH + xMPC + xMPN + xMPL - 2600 yPM <= 0 \Provo Mexican Cuisine
xIPH + xIPC + xIPN + xIPL - 2600 yPI <= 0 \Provo Italian Gourmet
   xADH + xMDH + xIDH + xADC + xMDC + xIDC \Dayton total production
 + xADN + xMDN + xIDN + xADL + xMDL + xIDL <= 2600
   xAPH + xMPH + xIPH + xAPC + xMPC + xIPC \Provo total production
 + xAPN + xMPN + xIPN + xAPL + xMPL + xIPL <= 2600

Bounds

\default is >= 0

General \non-negative product quantity

xADH xADC xADN xADL xAPH xAPC xAPN xAPL
xMDH xMDC xMDN xMDL xMPH xMPC xMPN xMPL
xIDH xIDC xIDN xIDL xIPH xIPC xIPN xIPL

Binary \a frozen dinner line is either produced or not produced in a plant

yDA yDM yDI yPA yPM yPI

End
```

The solution found by Gurobi Optimizer is as follows. Speedy Gourmet should produce all three lines of frozen dinner in the Dayton plant, but only Asian Cuisine and Mexican Cuisine in the Provo plant. The Dayton plant should produce 900,000 lb of Asian Cuisine; supplying 500,000 lb to the Philadelphia distribution center, 350,000 lb to the Chicago distribution center, and 50,000 lb to the New Orleans distribution center. The Provo plant should produce 500,000 lb of Asia Cuisine; supplying 100,000 lb to the New Orleans distribution center and 400,000 lb to the Las Vegas distribution center. The Dayton plant should produce 650,000 lb of Mexican cuisine; supplying 180,000 lb to the Philadelphia distribution center, 170,000 lb to the Chicago distribution center, and 300,000 lb to the New Orleans distribution center. The Provo plant should produce 450,000 lb of Mexican Cuisine to be supplied to the Las Vegas distribution center. The Dayton plant should produce 1,050,000 lb of Italian Gourmet; supplying 320,000 lb to the Philadelphia distribution center, 380,000 lb to the Chicago distribution center, 250,000 lb to the New Orleans distribution center, and 200,000 lb to the Las Vegas distribution center. The Dayton plant is operating at its full capacity of 2,600,000 lb of frozen food per year, whereas the Provo plant produces only 950,000 lb of frozen food. The sum of overhead cost and transportation cost is $1,045,150.

Case Study 5.5.4

QPF is a pet food manufacturer based in Cincinnati, Ohio. It has 10 major markets, namely, New York, Washington D.C., Philadelphia, Boston, Chicago, Detroit, Cleveland, Atlanta, St. Louis, and Memphis. The distances (in miles) from Cincinnati to these markets and the monthly demand (in thousand pounds) of the markets are shown in Table 5.7. The product of distance and demand is also shown in the weight-mile column in Table 5.7. QPF has decided to use a direct shipping strategy and identified three local trucking companies, Ada, Baxton, and Corbin, to transport its products. These companies charge a fixed cost (overhead) for each route per month plus a variable weight-mile cost (direct cost). Due to the availability of trucks, the monthly weight-mile capacity of these companies is limited. The information is shown in Table 5.8. Due to business strategy consideration, QPF has decided to assign no more than

TABLE 5.7

Market Distance (in Miles) and Demand (in Thousand lb)

Market	Distance	Demand	Weight-Mile
New York	580	70	40,600
Washington D.C.	520	65	33,800
Philadelphia	570	50	28,500
Boston	900	50	45,000
Chicago	300	90	27,000
Detroit	260	30	7,800
Cleveland	250	30	7,500
Atlanta	460	60	27,600
St. Louis	360	25	9,000
Memphis	490	20	9,800

TABLE 5.8

Trucking Company Cost and Capacity Information

Company	Fixed Cost (per Route)	Variable Cost (per Thousand lb × Mile)	Capacity (Thousand lb × Mile)
Ada	$25,000	$1.00	150,000
Baxton	$22,000	$1.15	200,000
Corbin	$20,000	$1.25	180.000

three routes to Ada. How should QPF assign the routes to the trucking companies?

We have 10 routes that need to be assigned to three trucking companies. The objective is to minimize the total transportation cost, subject to business agreement and trucking company capacity constraints. Again, this problem can be formulated as a mixed integer programming model. Let the markets be denoted from 0 to 9 and the trucking companies be denoted A, B, and C. We have 30 decision variables denoted xA0 (assign the New York route to Ada), xA1 (assign the Washington D.C. route to Ada), …, and xC9 (assign the Memphis route to Corbin). In addition to the business agreement and trucking company capacity constraints, our model needs to make sure that each of the 10 routes is assigned to one and only one trucking company. The Gurobi model for solving this problem is shown as follows:

```
Minimize

\minimize total transportation cost

  25000 xA0 + 25000 xA1 + 25000 xA2 + 25000 xA3 + 25000 xA4
+ 25000 xA5 + 25000 xA6 + 25000 xA7 + 25000 xA8 + 25000 xA9 \Ada overhead
+ 22000 xB0 + 22000 xB1 + 22000 xB2 + 22000 xB3 + 22000 xB4
+ 22000 xB5 + 22000 xB6 + 22000 xB7 + 22000 xB8 + 22000 xB9 \Baxton overhead
+ 20000 xC0 + 20000 xC1 + 20000 xC2 + 20000 xC3 + 20000 xC4
+ 20000 xC5 + 20000 xC6 + 20000 xC7 + 20000 xC8 + 20000 xC9 \Corbin overhead
+ 40600 xA0 + 46690 xB0 + 50750 xC0 \New York direct cost
+ 33800 xA1 + 38870 xB1 + 42250 xC1 \Washington D.C. direct cost
+ 28500 xA2 + 32775 xB2 + 35635 xC2 \Philadelphia direct cost
+ 45000 xA3 + 51750 xB3 + 56250 xC3 \Boston direct cost
+ 27000 xA4 + 31050 xB4 + 33750 xC4 \Chicago direct cost
+  7800 xA5 +  8970 xB5 +  9750 xC5 \Detroit direct cost
+  7500 xA6 +  8625 xB6 +  9375 xC6 \Cleveland direct cost
+ 27600 xA7 + 31740 xB7 + 34500 xC7 \Atlanta direct cost
+  9000 xA8 + 10350 xB8 + 11250 xC8 \St. Louis direct cost
+  9800 xA9 + 11270 xB9 + 12250 xC9 \Memphis direct cost

Subject To

xA0 + xA1 + xA2 + xA3 + xA4 + xA5 + xA6 + xA7 + xA8 + xA9 <= 3 \routes to Ada
   40600 xA0 + 33800 xA1 + 28500 xA2 + 45000 xA3 + 27000 xA4 \Ada capacity
+  7800 xA5 +  7500 xA6 + 27600 xA7 + 9000  xA8 + 9800  xA9 <= 150000
   40600 xB0 + 33800 xB1 + 28500 xB2 + 45000 xB3 + 27000 xB4 \Baxton capacity
+  7800 xB5 +  7500 xB6 + 27600 xB7 + 9000  xB8 + 9800  xB9 <= 200000
   40600 xC0 + 33800 xC1 + 28500 xC2 + 45000 xC3 + 27000 xC4 \Corbin capacity
+  7800 xC5 +  7500 xC6 + 27600 xC7 + 9000  xC8 + 9800  xC9 <= 180000
xA0 + xB0 + xC0 = 1 \one and only one trucking company for New York
xA1 + xB1 + xC1 = 1 \one and only one trucking company for Washington D.C.
xA2 + xB2 + xC2 = 1 \one and only one trucking company for Philadelphia
xA3 + xB3 + xC3 = 1 \one and only one trucking company for Boston
xA4 + xB4 + xC4 = 1 \one and only one trucking company for Chicago
xA5 + xB5 + xC5 = 1 \one and only one trucking company for Detroit
xA6 + xB6 + xC6 = 1 \one and only one trucking company for Cleveland
xA7 + xB7 + xC7 = 1 \one and only one trucking company for Atlanta
xA8 + xB8 + xC8 = 1 \one and only one trucking company for St. Louis
xA9 + xB9 + xC9 = 1 \one and only one trucking company for Memphis

Bounds

\default is >= 0

Binary \a route is either assigned or not assigned to a trucking company

xA0 xA1 xA2 xA3 xA4 xA5 xA6 xA7 xA8 xA9
xB0 xB1 xB2 xB3 xB4 xB5 xB6 xB7 xB8 xB9
xC0 xC1 xC2 xC3 xC4 xC5 xC6 xC7 xC8 xC9

End
```

The solution found by Gurobi Optimizer is to assign the New York, Washington D.C., and Boston routes to Ada; assign the Philadelphia, Chicago, and Atlanta routes to Baxton; and assign the Detroit, Cleveland, St. Louis, and Memphis routes to Corbin. The total monthly cost for the assignment is $478,590.

Problem: Dragon Furniture Enterprise

Dragon Furniture Enterprise (DFE) is a manufacturer of Chinese redwood dining sets based in Shenzhen, Guangdong Province. It has been selling its products to the U.S. market for several years through various wholesalers. It has four major markets in the United States, namely, Los Angeles, San Francisco, Houston, and New York. The distances among these markets are shown in Table 5.9. Weekly demand follows a normal distribution with details shown in Table 5.10. The demand between Los Angeles and San Francisco is correlated with a correlation coefficient of 0.5. All other demands are independent. DFE aims to achieve a cycle service level of 97%.

To improve its profit margin, DFE is now considering building its own distribution centers in the United States. Building a distribution center requires both capital investment and annual operating costs. The capital investment is $800,000 + 100x$ and the operating cost is $100,000 + 5x$, where x is the capacity (number of dining sets) of the distribution center. The capital investment is depreciated evenly over 30 years. The inventory holding cost for a dining set is $500 per year.

TABLE 5.9

Distances (in Miles) among DFE's Four Major Markets

	Los Angeles	San Francisco	Houston	New York
Los Angeles	—	400	1600	2800
San Francisco	—	—	1950	3000
Houston	—	—	—	1650
New York	—	—	—	—

TABLE 5.10

Weekly Demand of Redwood Dining Sets

	Market			
Demand	Los Angeles	San Francisco	Houston	New York
Mean	700	650	600	1200
Standard deviation	100	100	80	200

DFE is considering two distribution center options. The first option is to build a single distribution center in one of the markets. Dining sets will be shipped from Shenzhen to this distribution center once every 10 weeks by ocean liners. Each shipment takes 4 weeks and costs $200,000 to either Los Angeles or San Francisco; takes 5 weeks and costs $220,000 to Houston; and takes 6 weeks and costs $250,000 to New York. Customers in the local market pick up dining sets in DFE's distribution center. For customers in the other markets, DFE has to deliver the dining sets using a trucking company. This outbound transportation cost is estimated at $50 per dining set per 1000 miles travelled.

The second option is to build a distribution center in each market. Dining sets will be shipped from Shenzhen to the Port of Los Angeles and then loaded to a train and shipped to each distribution center. Shipping to the distribution centers at Los Angeles and San Francisco is done overnight, whereas shipping to the distribution centers at Houston and New York takes 1 week. When using this option, DFE incurs inbound transportation cost to its distribution centers, in addition to shipping from Shenzhen to the Port of Los Angeles. Due to the economies of scale and the use of train, the transportation cost is only $5 per dining set per thousand miles travelled. On the other hand, DFE does not incur outbound transportation cost because all customers pick up the dining sets from its distribution centers.

If DFE uses the first option, where should the distribution center be built? Is the second option better than the first option?

Exercises

5.1   TSS Manufacturing Company, an Ohio-based automobile parts supplier, received orders from several automobile manufacturers to be delivered within a month. The orders are as follows: 600, 750, 560, and 820 units to be delivered to the assembly plants in Georgetown (Kentucky), Marysville (Ohio), Toledo (Ohio), and Warren (Michigan), respectively. TSS has three manufacturing plants in Cincinnati, Akron, and Findlay, respectively. The unit production cost is the same in all three plants. The monthly production capacities of these plants are 2000, 1000, and 1000 units, respectively. The transportation cost is $.25 per unit per mile. The transportation distance (in miles) is shown as follows:

	Georgetown	Marysville	Toledo	Warren
Cincinnati	70	120	200	280
Akron	300	130	140	210
Findlay	230	70	50	120

How should TSS schedule its production and delivery in order to minimize transportation cost?

5.2 Consider a supply chain in which a number of distribution centers supply several markets. The goal is to identify distribution center locations as well as quantities shipped from distribution centers to markets that minimize the total costs. One distribution center can supply multiple markets. However, a market can only be supplied by a distribution center. Formulate this constrained optimization problem using the following notations:

a: number of markets (demand points)
b: number of potential distribution center locations
D_i: annual demand from market i, $i = 1, 2, …, a$
K_j: potential annual capacity of distribution center at site j, $j = 1, 2, …, b$
F_j: annual cost of locating a distribution center at site j
c_{ji}: cost of shipping one unit from distribution center j to market i

Hint: a constrained optimization problem is not necessarily a linear programming or mixed integer programming problem.

5.3 Golden Spice Inc. was established 20 years ago in Los Angeles. It imports spice from Southeast Asia in bulk and repackages it for selling to various supermarkets and convenience stores. Its business has grown substantially and the company established regional distribution centers in Philadelphia, Chicago, and Orlando. The markets served by the four distribution centers are roughly equal in size. The monthly demand in each distribution center is independent and follows a normal distribution with a mean of 10,000 lb and a standard deviation of 6,000 lb. The total annual inbound and outbound transportation costs are $220,000 and $480,000, respectively. The inventory holding cost is $5 per lb/year. In recent years, due to increased competition, Golden Spice has to increase its CSL to 99%. It was not able to shorten its replenishment lead time, which remains at 4 months. As a result, the company's inventory holding cost has increased significantly. A supply chain consultant suggested that the company should close its regional distribution center so inventory can be aggregated in its Los Angeles main distribution center. This will not only reduce inventory holding cost but also reduce total inbound transportation cost by 50%. In addition, the annual distribution center operating cost can be reduced by $300,000. To maintain the same level of responsiveness, the consultant suggested using a combination of highway and air for outbound transportation. It was estimated that by doing so the outbound transportation cost will increase by between 140% and 180%. Assuming there is no change in Golden Spice's replenishment policy, should the company close its regional distribution centers?

5.4 Refer to Case 5.1, suppose the time and cost for transporting closet organizers from New Jersey to the warehouse in San Francisco remains the same. However, the time and cost for transporting closet organizers from New Jersey to the warehouse in Los Angeles is as follows:

- By railroad, 2 weeks with a cost of $0.42 per closet organizer.
- By highway, 1 week with a cost of $.95 per closet organizer.

Should the warehouse be established in San Francisco or Los Angeles?

5.5   A city has three police stations, A, B, and C. The number of police officers in the stations is 3, 6, and 7, respectively. One day, the central dispatching center received 10 emergency calls. The travel time (in minutes) between the police stations and the emergency locations are shown as follows:

	1	2	3	4	5	6	7	8	9	10
A	3.7	8.6	2.8	1.9	1.0	6.9	7.9	4.1	7.3	2.9
B	1.3	6.1	6.8	5.4	3.7	5.3	5.6	3.5	4.4	6.7
C	5.9	3.7	2.2	3.4	5.2	2.6	3.2	3.6	3.5	2.1

How should the police officers be dispatched in order to minimize the total response time?

References

Chopra, S. and P. Meindl. 2010. *Supply Chain Management: Strategy, Planning, and Operation*, 4th edn. Upper Saddle River, NJ: Pearson Prentice Hall.

Ferdows, K. 1997. Making the most of foreign factories. *Harvard Business Review* **75**: 73–88.

Flood, M. M. 1956. The traveling-salesman problem. *Operations Research* **4**: 61–75.

Goldreich, O. 2010. *P, NP, and NP-Completeness: The Basics of Computational Complexity*. Cambridge, MA: Cambridge University Press.

Lodi, A., S. Martello, and M. Nonaci. 2002. Two-dimensional packing problems: A survey. *European Journal of Operational Research* **141**: 241–252.

Stank, T. P. and T. J. Goldsby. 2000. A framework for transportation decision making in an integrated supply chain. *Supply Chain Management: An International Journal* **5**: 71–77.

6

Supplier Selection Methodology

6.1 Overview

In Chapters 2 through 5, we presented quantitative analysis methods to help a company better understand customer demand, plan production and manage inventory to satisfy demand, and design a distribution network to efficiently deliver products to customers. The company cannot accomplish all of these tasks alone. Rather, it has to rely on its suppliers to provide certain products and services. The process by which a company acquires raw materials, components, products, or services from suppliers is called *Purchasing. Sourcing* is the business process required to purchase goods and services. According to Krajewski and Ritzman (2001), the percentage of sales revenues spent on purchased materials varies from more than 80% in the petroleum refining industry to 25% in the pharmaceutical industry. Therefore, sourcing is an important issue that an original equipment manufacturer (OEM) must address. It has a significant impact on supply chain profitability.

To satisfy customers' demand of higher quality at lower price, OEMs must use suppliers that can deliver required raw materials and components at a high-quality level with low cost. In addition, because of shortened product life cycle, OEMs and suppliers need to develop strategic partnerships so they can adapt to a rapidly changing market. Furthermore, with rising consumerism and the concern about the environment, more and more OEMs are consciously building a consumer- and environmental-friendly image, partially reflected in their supply chain practices. The sourcing process includes the following steps (Chopra and Meindl 2010):

- *Supplier scoring and assessment*: In this step, suppliers are rated based on their impact on supply chain performance. The performance measures should cover a variety of aspects including price, delivery lead time, reliability, quality, technical capability, and financial stability.

- *Supplier selection and contract negotiation*: Potential suppliers are identified based on the output from the supplier scoring and assessment step. Contracts are then negotiated. A good contract should be designed to increase supply chain profitability that benefits both the manufacturer and the supplier.

- *Design collaboration*: Design collaboration allows the manufacturer and the supplier to combine their resource and technical capability to produce more cost-effective product designs. It also ensures that any design changes are communicated effectively to all parties along the supply chain.

- *Procurement*: The goal of procurement is to enable orders to be placed and delivered on schedule at the lowest possible overall cost.

- *Sourcing planning and analysis*: In this step, the manufacturer analyzes spending across various suppliers and component categories to identify opportunity to maximize supply chain profit.

In this chapter, we focus on issues related to supplier selection, namely, performance measures for supplier scoring and assessment, the methodology used for supplier selection, and contracts to increase supply chain profitability.

6.2 Supplier Performance Measures

When measuring the performance of suppliers, cost and quality have been the most dominant factors, along with on-time delivery and flexibility. Literature in the late 1970s and early 1980s showed heavy emphasis on cost. In the early 1990s, cycle time and customer responsiveness were added. In the late 1990s, researchers realized the importance of flexibility. In recent years, environmental safety became a key issue among industrialized nations. This gives rise to the concept of green supply chain. Performance in this area is measured using various metrics depending on product properties, recycling, waste/hazardous emission, and resource usage. The trend is shifting toward developing more exhaustive and detailed performance metrics in a systematic way.

Built on previous work in the literature, Huang and Kaskar (2007) presented a comprehensive set of supplier performance metrics. The metrics are grouped into seven categories: (1) reliability, (2) responsiveness, (3) flexibility, (4) cost and financial, (5) asset and infrastructure, (6) safety, and (7) environmental. These seven categories of performance metrics are further organized into three tracks, that is, product related, supplier related, and society related, for easier user configuration. Metrics belong to the five categories related to product type and supplier type are further partitioned to match OEM/supplier integration level. There are three levels of OEM/supplier integration, namely, no integration, operational integration, and strategic partnership. Metrics that

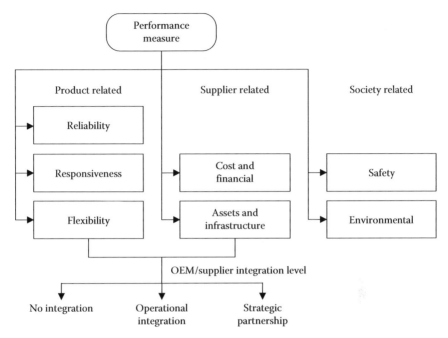

FIGURE 6.1
Hierarchy of supplier performance measure. Reprinted from *Int. J. Prod. Econ.*, 105(2), Huang, S.H., and Keskar, H., Comprehensive and configurable metrics for supplier selection, 510–523, Copyright 2007, with permission from Elsevier.

are measured when the OEM receives the ordered component are mapped to the no integration level. Metrics that need to be measured at the supplier's facility are mapped to the level of operational integration. Metrics mapped to the level of strategic partnership are those considered proprietary and hence are available only when the OEM and the supplier have formed a strategic partnership. Safety and environmental metrics are society related and are not considered proprietary. Therefore, they are not partitioned into different levels. The supplier performance measure hierarchy is shown in Figure 6.1.

A total of 101 metrics were collected, evaluated, and categorized. There are 19 reliability metrics, summarized in Table 6.1. These metrics are criteria regarding the performance of a supplier in delivering the ordered components to the right place, at the agreed upon time, in the required condition and packaging, and in the required quantity. There are 13 responsiveness metrics, summarized in Table 6.2. These metrics are criteria related to the velocity at which a supplier provides products to the customer. There are 14 flexibility metrics, summarized in Table 6.3. These metrics are criteria regarding the agility of a supplier in responding to OEM demand changes. There are 17 cost and financial metrics, summarized in Table 6.4. These metrics are criteria regarding cost and financial aspects of procuring from supplier. There are 19 asset and infrastructure metrics, summarized in Table 6.5. These metrics are criteria

TABLE 6.1

Reliability Metrics

Metrics	Definition	Integration Level
% Orders received damage free	Number of orders received damage free divided by total number of orders processed in measurement time	No integration
% Orders received complete	Number of orders received complete divided by total number of orders processed in measurement time	No integration
% Orders received on time to commit date	Number of orders received on time to commit date divided by total number of orders processed in measurement time	No integration
% Orders received on time to required date	Number of orders received on time to required date divided by total number of orders processed in measurement time	No integration
% Orders received defect free	Number of orders received defect free divided by total number of orders processed in measurement time	No integration
% Orders received with correct shipping docs	Number of orders received with correct shipping docs divided by total number of orders processed in measurement time	No integration
% Orders scheduled to customer request date	Percentage of orders whose delivery is scheduled within an agreed time frame of the customer's requested delivery date	No integration
% Faultless installations	Number of faultless installations divided by total number of units installed	No integration
Scrap expenses	Expenses incurred from material failing outside of specifications and processing characteristics that make rework impractical as percentage of total production cost	No integration
% Short to manufacturing schedule	Number of orders produced exceeding the manufacturing schedule divided by total number of orders produced in measurement time	Operational integration
Fill rate	The percentage of ship-from-stock orders shipped within 24 h of order receipt	Operational integration
Ratio of actual to theoretical cycle time	Ratio of measured time required for completion of set of tasks divided by sum of the time required to complete each task based on rated efficiency of the machinery and labor operations	Operational integration
Yields during manufacturing	Ratio of usable output from a process to its input	Operational integration
% Errors during release of finished product	Number of errors in release of finished products divided by total number of products released during measurement period	Operational integration
Inventory accuracy	The absolute value of the sum of the variance between physical inventory and perpetual inventory	Operational integration

TABLE 6.1 (continued)

Reliability Metrics

Metrics	Definition	Integration Level
Order consolidation profile	The activities associated with filling a customer order by bringing together in one physical place all of the line items ordered by the customer	Operational integration
Average days per engineering change	Total number of days each engineering change impacts the delivery date divided by the total number of changes	Operational integration
Incoming material quality control	Quality assurance procedures, control over quality of incoming material at supplier and quality improvement perspective toward supplier's suppliers	Strategic partnership
In process failure rate	The percentage of work-in-process that is not completed, i.e., 1 minus the percentage of completed work-in-process units	Strategic partnership

Source: *Int. J. Prod. Econ.*, 105(2), Huang, S.H., and Keskar, H., Comprehensive and configurable metrics for supplier selection, 510–523, Copyright 2007, with permission from Elsevier.

regarding the effectiveness of supplier in managing assets to support OEM demand. There are five safety metrics, summarized in Table 6.6. These metrics are criteria regarding occupational safety at the supplier's facility. There are 14 environmental metrics, summarized in Table 6.7. These metrics are criteria regarding a supplier's effort in pursuing environmental friendly production.

The categorization is by no means rigid and the performance metrics are not exhaustive. When evaluating potential suppliers, a company will not need to use all of these metrics. Rather, it should first develop a supply chain strategy based on product characteristics, as discussed in Chapter 1. It then decides on what metric categories should be used to support its supply chain strategy. Finally, a number of specific metrics are selected to evaluate its suppliers.

6.3 Supplier Selection Methods

Generally speaking, the supplier selection problem can be described as follows. A manufacturing company A produces m products. For each product P_i (product index $i = 1, 2, ..., m$), it may consist of n_i major components, which need to be outsourced (company A might have capacities to produce the other components by itself). For each outsourced component C_{ij} (component index $j = 1, 2, ... n_i$), there are k_{ij} potential suppliers to choose from. Each potential supplier S_{ijx} (supplier index $x = 1, 2, ..., k_{ij}$) has a known production capacity R_{ijx}. According to the production plan, company A will purchase T_{ij} units of component C_{ij} from one or more suppliers out of the whole set

TABLE 6.2

Responsiveness Metrics

Metrics	Definition	Integration Level
Published delivery cycle time	Typical standard lead time after receipt of order currently published to customers by the sales organization.	No integration
Order fulfillment lead time	The average actual lead time consistently achieved, from customer signature/ authorization to order receipt, order receipt to order entry complete, order entry complete to start-build, start-build to order ready for shipment, order ready for shipment to customer receipt of order, and customer receipt of order to installation complete.	No integration
Return product velocity	Average time required for process of returning the defective, incomplete, or damaged orders and reshipping of the order to customer.	No integration
Average release cycle of changes	Cycle time for implementing change notices divided by total number of changes.	Operational integration
Total build cycle time	Total build time is the average time for build-to-stock or configure-to-order products from when production begins on the released work order until the build is completed and unit deemed shippable.	Operational integration
Package cycle time	The total time required to perform a series of activities that containerize completed products for storage or sale to end-users.	Operational integration
Product release process cycle time	Total time required to perform post-production documentation, testing, or certification required prior to delivery of finished product to customer.	Operational integration
Installation cycle time	Total time required to prepare and install the product at customer site with full functional commencement.	Operational integration
Sourced/in process product requisition cycle time	The time required to provide manufacturing with a needed component, service, or additive from the time of requisition to the time of delivery.	Strategic partnership
Product/grade change over time	Average time required to change from one product or grade to another product or grade.	Strategic partnership
Intra production re-plan cycle time	Time between the acceptance of a regenerated forecast by the end-product producing location and the reflection of the revised plan in the master production schedule of all the affected plants, excluding external vendors.	Strategic partnership

TABLE 6.2 (continued)

Responsiveness Metrics

Metrics	Definition	Integration Level
Quarantine/hold time	Average time for setting aside of items from availability for use or sale until all required quality tests have been performed and conformance certified.	Strategic partnership
Production engineering cycle time	Average time required for generation and delivery of final drawings, specifications, formulas, part programs, etc. In general, preliminary engineering work done as part of the quotation process.	Strategic partnership

Source: *Int. J. Prod. Econ.*, 105(2), Huang, S.H., and Keskar, H., Comprehensive and configurable metrics for supplier selection, 510–523, Copyright 2007, with permission from Elsevier.

of potential suppliers for component C_{ij} based on company A's predefined supplier selection criteria considering each supplier's production capacity. In summary, company A will make decision:

- To choose preferred suppliers for various outsourced components that meet its supplier selection criteria
- To order various quantities from the chosen suppliers to meet its production plan

As presented in Section 6.2, there are a large number of supplier selection metrics. A company generally selects at least a handful of metrics from different categories. Therefore, supplier selection is a multicriteria decision-making (MCDM) problem. The metrics could be quantitative (e.g., order fulfillment lead time) or qualitative (e.g., cultural similarity). To handle this problem, two MDCM approaches are commonly used, namely, analytical hierarchy process (AHP) and multiattribute utility theory (MAUT).

6.3.1 Analytical Hierarchy Process Approach

Wang et al. (2004) proposed an integrated supplier selection approach based on AHP and preemptive goal programming. AHP is a decision-making tool that can help describe the general decision operation by decomposing a complex problem into a multilevel hierarchical structure of objectives, criteria, subcriteria, and alternatives (Satty 1980). A general configuration of AHP-based hierarchy for supplier selection is illustrated in Figure 6.2. The steps used to determine supplier preference are summarized as follows:

Step 1: *Decompose problem*—The objective of supplier selection is to find a supplier with the highest overall preference rating. This level 1 objective is decomposed into several categories of level 2 criteria, namely, reliability, flexibility, responsiveness, cost and financial, asset and infrastructure, safety, and environmental.

TABLE 6.3

Flexibility Metrics

Metrics	Definition	Level
Time for expediting delivery and transfer process	Expediting cycle time for delivery and transfer process compared to the standard cycle time for the delivery and transfer process	No integration
Cost of expediting delivery and transfer process	Additive cost required to expedite the delivery and transfer process by the supplier	No integration
Ability to augment return capacity rapidly	Appropriation of return resources and assets to meet anticipated as well as unanticipated return requirements	Operational integration
ECO cycle time	The total time required from request for change from customer, engineering, production, or quality control to revise a blueprint or design released by engineering, and implement the change within the Make operation	Operational integration
Upside order flexibility	Number of days required to achieve an unplanned sustainable 20% increase in orders	Strategic partnership
Downside order flexibility	Percentage order reduction sustainable at 30 days prior to shipping with no inventory or cost penalties	Strategic partnership
Upside production flexibility	Number of days required to achieve an unplanned sustainable 20% increase in production	Strategic partnership
Downside production flexibility	The percentage production reduction sustainable at 30 days prior to delivery with no inventory or cost penalties	Strategic partnership
Upside delivery flexibility	Number of days required to achieve an unplanned sustainable 20% increase in deliveries	Strategic partnership
Downside delivery flexibility	Percentage delivery reduction sustainable at 30 days prior to delivery with no inventory or cost penalties	Strategic partnership
Upside installation flexibility	Number of days required to achieve an unplanned sustainable 20% increase in installations	Strategic partnership
Downside installation flexibility	Percentage installation reduction sustainable at 30 days prior to installing with no inventory or cost penalties	Strategic partnership
Upside shipment flexibility	Number of days required to achieve an unplanned sustainable 20% increase in shipments	Strategic partnership
Downside shipment flexibility	Percentage shipment reduction sustainable at 30 days prior to shipping with no inventory or cost penalties	Strategic partnership

Source: *Int. J. Prod. Econ.*, 105(2), Huang, S.H., and Keskar, H., Comprehensive and configurable metrics for supplier selection, 510–523, Copyright 2007, with permission from Elsevier.

TABLE 6.4

Cost and Financial Metrics

Metrics	Definition	Integration Level
Inventory turns	Total cost of goods sold divided by value of inventory carried throughout the measurement period	No integration
Payment terms	Suitability of terms and conditions regarding payment of invoices, open accounts, sight drafts, credit letter, and payment schedule	No integration
Return policy	Suitability of policies regarding return of the defective, damaged, or incomplete orders	No integration
Warranty costs	Warranty costs include materials, labor, and problem diagnosis for product defects	No integration
Landed cost	Final cost including the cost of components/order, shipping cost, duties, broker fees, custom fees, qualification fees, etc., required to be paid per component/order	No integration
Discount rate	Suitability of discount scheme implemented on payment of invoices within time frame	No integration
Foreign exchange rate fluctuation	Fluctuation in frequency and range of the currency exchange rate between the two countries	No integration
Financial stability	Indicator of excessive asset price volatility, the unusual drying-up of liquidity, interruptions in the operation of payment systems, excessive credit rationing etc.	Operational integration
Packaging cost	Cost of series of activities that containerize completed products for storage or sale to end-users	Operational integration
Inventory carrying cost	Inventory carrying costs are the sum of opportunity cost, shrinkage, insurance and taxes, total obsolescence for raw material, WIP, and finished goods inventory, channel obsolescence, and field sample obsolescence	Operational integration
Order fulfillment costs	Includes costs for processing the order, allocating inventory, ordering from the internal or external supplier, scheduling the shipment, reporting order status, and initiating shipment	Operational integration
Freight	Costs of transporting component from supplier facility to customer facility	Operational integration
Local price control	Suitability of price control and counter trade policies due to country government policies and local government rules and regulations	Operational integration
Tariffs and custom duties	Custom duties/tariffs imposed by importing country on goods and services imported from particular country	Operational integration

(*continued*)

TABLE 6.4 (continued)

Cost and Financial Metrics

Metrics	Definition	Integration Level
Value-added productivity	Value added per employee is calculated as total product revenue less total material purchases divided by total employment (in full-time equivalents)	Strategic partnership
Release cost per unit	Cost involved in post-production documentation, testing, or certification required prior to delivery of finished product to customer	Strategic partnership
Cost reduction trend	Average change in operating costs during the measurement period	Strategic partnership

Source: Int. J. Prod. Econ., 105(2), Huang, S.H., and Keskar, H., Comprehensive and configurable metrics for supplier selection, 510–523, Copyright 2007, with permission from Elsevier.

Step 2: *Define criteria for supplier selection*—Under each level 2 performance category, appropriate metrics are determined based on the company's supply chain strategy. These metrics form the level 3 supplier selection subcriteria.

Step 3: *Design the hierarchy*—The hierarchy consists of the level 1 objective, level 2 performance criteria, level 3 performance subcriteria (specific metrics selected from Tables 6.1 through 6.7), and level 4 decision alternatives (potential suppliers).

Step 4: *Perform pair-wise comparison and prioritization*—Once the problem has been decomposed and the hierarchy constructed, prioritization procedure starts in order to determine the relative importance of the elements within each level. The pair-wise judgment starts from level 2 and continues on to level 3. Based on product characteristics and corresponding supply chain strategies, the relative importance of the performance categories and performance metrics are determined by experienced managers.

Step 5: *Rate the potential suppliers*—Similar to step 4, the potential suppliers are compared pair-wisely. The comparison is made with respect to each of the level 3 performance metrics.

Step 6: *Calculate the weights of the performance category, performance metrics, and potential suppliers*—Generally, given a pair-wise comparison matrix, the priority weights for each attribute can be calculated based on standard methods provided by Satty (1980).

Step 7: *Compute the overall score of each supplier*—By integrating the assigned weights of the attributes, the final supplier scores are determined.

Step 8: *Make overall decision*—The supplier (decision alternative) that has the highest rating is the best choice. If there are no capacity constraints, this supplier is chosen to satisfy all the product demand. Otherwise, suppliers with lower ratings need to be considered.

TABLE 6.5

Assets and Infrastructure Metrics

Metrics	Definition	Integration Level
Political stability	Political stability and relations with the exporting country	No integration
Labor stability	Labor turn over during the measurement period within various employee categories	No integration
Asset turns	Total gross product revenue divided by total net assets	No integration
Company size	Indicator of various factors such as facility size, area, work force strength, turnover, capacity, etc.	No integration
Quality system certification/ assessment	Quality certifications acquired and performance on conformance audits during measurement period	No integration
Strategic fit	Compatibility of long term planning in regards to expansion plans, area of concentration, interest in collaborating, etc.	No integration
Negotiability	Negotiation flexibility with regards to cost, payment terms, return policies, and other similar terms and conditions in supplier-buyer contract	No integration
Legal claims	Pending or filed legal claims against the supplier	No integration
Critical process subcontracting	Percentage of critical process subcontracted by supplier	Operational integration
Inventory days of supply	Total gross value of inventory at standard cost before reserves for excess and obsolescence including inventory on company books only excluding future liabilities	Operational integration
Capacity utilization	A measure of how intensively a resource is being used to produce a good or service. Some factors that should be considered are internal manufacturing capacity, constraining processes, direct labor availability and key components/materials availability	Operational integration
Management outlook and functional compatibility	Degree of alignment in future plans, management policies, competitive strategies, and match between various functions across the supplier organization	Operational integration
Ethical standards	Compatibility of ethical standards practiced at supplier end	Operational integration
Designing capabilities	Capabilities regarding conceptualizing, designing, drafting, and prototyping of new product requirements	Strategic partnership
Development capabilities	Capabilities regarding development of manufacturing processes, trial runs, quality assurance for new product design	Strategic partnership
EDI capabilities	Capabilities and infrastructure regarding electronic data transfer at supplier facility for effective communication	Strategic partnership

(*continued*)

TABLE 6.5 (continued)

Assets and Infrastructure Metrics

Metrics	Definition	Integration Level
Manufacturing/ process capabilities	Capabilities in areas of machining, manufacturing, assembly, special purpose machines and equipment, etc., in line with product requirements	Strategic partnership
Customer concentration	Percentage share of sales from the supplier as compared to other buyers	Strategic partnership
Cultural similarity	Cultural and language barriers between buyer and supplier	Strategic partnership

Source: *Int. J. Prod. Econ.*, 105(2), Huang, S.H., and Keskar, H., Comprehensive and configurable metrics for supplier selection, 510–523, Copyright 2007, with permission from Elsevier.

TABLE 6.6

Safety Metrics

Metrics	Definition
Number of lost time accidents	Number of accidents per million working hours resulting in lost time
Recordable incident rate	OSHA or equivalent recordable incident rate per 100 employees
Dollars spent in worker compensation	Total dollar amount spent on worker compensation due to work-related injury during the measurement period
Safety training	Procedures and practices regarding safety training and level of awareness
Safety audits	Feedback of safety procedures through review audit

Source: *Int. J. Prod. Econ.*, 105(2), Huang, S.H., and Keskar, H., Comprehensive and configurable metrics for supplier selection, 510–523, Copyright 2007, with permission from Elsevier.

Wang et al. (2005) used preemptive goal programming to make the overall decision. Other optimization approaches can also be used to determine order quantity. The simplest method is to proceed from the supplier with the highest preference rating to the one with the second highest preference rating and so forth until demand is satisfied. The following example was provided by Wang et al. (2005) to illustrate the AHP approach to supplier selection.

GW Inc. is a hypothetical car manufacturer that can produce various functional components, such as engine, body, and glass components. Only three components need to be outsourced, namely, tires, electronics, and peripherals. To determine supplier preference, the company decided to combine flexibility and responsiveness as a single criterion. In addition, reliability, cost, and assets are identified as the other criteria to evaluate suppliers. Specific performance metrics are then selected as follows:

- Reliability: delivery performance (DR1), fill rate (DR2), order fulfillment lead time (DR3), and perfect order fulfillment (DR4)
- Flexibility and responsiveness: supply chain response time (FR1), production flexibility (FR2)

TABLE 6.7

Environmental Metrics

Metrics	Definition
Conventional pollutants released to water	Average volume of conventional pollutants (suspended solids, biological oxygen demand, fecal coliform bacteria, pH, and oil and grease) released per day during measurement period
Ambient air releases	Average volume in ppmv of ambient air released per day during measurement period
Hazardous/ nonhazardous waste	Average volume of hazardous/nonhazardous waste released per day during measurement period
Chemical releases	Average volume of chemical released per day during measurement period
Global warming gases	Average volume in ppmv of global warming gas (carbon dioxide, methane) released per day during measurement period
Ozone depleting chemicals	Average volume of ozone depleting chemicals released per day during measurement period
Bio accumulative pollutants	Average volume of bio accumulative pollutants released per day during measurement period
Indoor environmental releases	Average volume of indoor environmental waste released per day during measurement period
Resource consumption (material, energy, water)	Resource consumption in terms of material, energy and water during the measurement period
Non renewable resource consumption	Resources not renewable in 200 years (fossil fuels minerals etc.) consumed in terms during the measurement period
Recycled content	Percentage of materials that can be recovered from the solid waste stream, either during the manufacturing process or after consumer use
Product disassembly potential	Ease with which a product can be disassembled for maintenance, replacement or recycling
Product durability	Measure of useful life of the product
Component reusability	Percentage of reusable components in total number of components in the product and their frequency of reusability

Source: *Int. J. Prod. Econ.*, 105(2), Huang, S.H., and Keskar, H., Comprehensive and configurable metrics for supplier selection, 510–523, Copyright 2007, with permission from Elsevier.

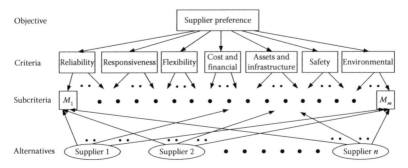

FIGURE 6.2
AHP-based hierarchy for supplier selection.

- Cost: total logistics management cost (CT1), value-added productivity (CT2), warranty cost/returns processing cost (CT3)
- Assets: cash-to-cash cycle time (AT1), inventory days of supply (AT2), asset turns (AT3)

It is very difficult for a supplier to achieve excellent performance in all of the selected metrics. Therefore, it is necessary for GW Inc. to prioritize the performance category based on component characteristics. Consider tires first. They are functional products whose demand is stable. Therefore, the focus should be on reducing costs, while flexibility and responsiveness are not so important. Now consider peripherals. To satisfy consumers' personal preference, GW Inc. plans to offer a large variety of combinations. In the mean time, new peripherals (such as navigation systems) appear due to rapid technology development. Therefore, peripherals are considered as innovative products whose demand is difficult to forecast. In this case, flexibility and responsiveness become the priority, while cost issues are secondary. Finally, electronics require mass customization, yet its aggregated demand is stable. Therefore, they are considered hybrid products where the concept of postponement can be used—differentiate late in the manufacturing process with short lead time items, and standardize long lead time items to be processed early in the supply chain. In this case, a more balanced view toward cost and flexibility requirements should be taken.

Company managers now need to prioritize the performance metrics based on the three different supply chain strategies. This is done by using pair-wise comparison with Satty's 1-9 scales. Note that pair-wise comparison involves subjective ratings. This process is carried out by a number of experts. AHP weights are then calculated and a consistency check is performed, which can be done using a software tool. The final AHP ratings for lean, agile, and hybrid supply chain strategies are shown in Table 6.8. One can see that the lean supply chain strategy has an emphasis on cost; the agile supply chain strategy has an emphasis on flexibility and responsiveness; whereas the hybrid supply chain strategy has a more balanced weighting on the four criteria.

For each of the three components, there are several potential suppliers that have been certified by the Quality Assurance Department of GW Inc. It is a common practice for manufacturing companies to use certain quality criteria (such as ISO 9000 certification) to identify a limited number of potential suppliers. This pre-selection process reduces the number of decision alternatives. It can be carried out based on strategic quality management practices, such as those presented by Tummala and Tang (1996). It is assumed that only three potential suppliers are qualified to supply each outsourced component, respectively. A_1, A_2, and A_3 represent the three potential suppliers for tires; A_4, A_5, and A_6 represent the three potential suppliers for electronics; and A_7, A_8, and A_9 represent the three potential suppliers for peripherals. Furthermore, A_1, A_4, and A_7 are assumed to be lean-oriented suppliers,

TABLE 6.8

AHP Rating for Different Supply Chain Strategy

Strategy	Objective	Reliability	Flexibility/ Responsiveness	Cost	Asset
Lean	DR 0.243	DR1 0.200	FR1 0.500	CT1 0.500	AT1 0.333
	FR 0.046	DR2 0.200	FR2 0.500	CT2 0.250	AT1 0.333
	CT 0.640	DR3 0.200		CT3 0.250	AT3 0.333
	AT 0.072	DR4 0.400			
Agile	DR 0.187	DR1 0.200	FR1 0.667	CT1 0.333	AT1 0.333
	FR 0.647	DR2 0.200	FR2 0.333	CT2 0.333	AT1 0.333
	CT 0.082	DR3 0.200		CT3 0.333	AT3 0.333
	AT 0.057	DR4 0.400			
Hybrid	DR 0.198	DR1 0.200	FR1 0.667	CT1 0.500	AT1 0.333
	FR 0.387	DR2 0.200	FR2 0.333	CT2 0.250	AT1 0.333
	CT 0.275	DR3 0.200		CT3 0.250	AT3 0.333
	AT 0.140	DR4 0.400			

Source: Modified from *Int. J. Prod. Econ.*, 91(1), Wang, G., Huang, S. H., and Dismukes, J.P., Product driven supply chain selection using integrated multi-criteria decision making methodology, 1–15, Copyright 2004, with permission from Elsevier.

A_2, A_5, and A_8 are assumed to be agile-oriented suppliers, and A_3, A_6, and A_9 are assumed to be hybrid-oriented suppliers. Table 6.9 lists the supplier information for each component and the OEM demand information.

A pair-wise comparison among the suppliers with respect to the 12 specific performance metrics is then carried out. Again this process is carried out by a number of experts, and the AHP ratings for suppliers A_1, A_2, and A_3 are shown

TABLE 6.9

Supplier and Demand Information for GW's Product

Component	Tires			Electronics			Peripherals		
Supplier	A_1	A_2	A_3	A_4	A_5	A_6	A_7	A_8	A_9
Capacity (units)	400	700	600	100	300	200	150	200	150
Defect rate (%)	1	3	2	1	3	2	1	3	2
Unit purchasing cost ($)	0.6	2.4	1.2	0.6	2.4	1.2	0.6	2.4	1.2
Fill rate (%)	80	90	98	80	90	98	80	90	98
OEM demand (units)		1000			250			250	
Max. acceptable defect rate (%)		2			2			2	

Source: *Int. J. Prod. Econ.*, 91(1), Wang, G., Huang, S. H., and Dismukes, J.P., Product driven supply chain selection using integrated multi-criteria decision making methodology, 1–15, Copyright 2004, with permission from Elsevier.

TABLE 6.10

AHP Ratings for Suppliers

	DR1	DR2	DR3	DR4	FR1	FR2	CT1	CT2	CT3	AT1	AT2	AT3
A_1	0.581	0.299	0.557	0.557	0.110	0.110	0.571	0.557	0.545	0.110	0.164	0.164
A_2	0.110	0.336	0.123	0.123	0.581	0.581	0.143	0.123	0.182	0.309	0.297	0.297
A_3	0.309	0.366	0.320	0.320	0.309	0.309	0.286	0.320	0.273	0.581	0.539	0.539

Source: *Int. J. Prod. Econ.*, 91(1), Wang, G., Huang, S. H., and Dismukes, J.P., Product driven
        supply chain selection using integrated multi-criteria decision making methodol-
        ogy, 1–15, Copyright 2004, with permission from Elsevier.

in Table 6.10. The final preference ratings for A_1, A_2, and A_3 are 0.499, 0.171, and 0.320, respectively. This means that A_1 is the most preferred, lean-oriented supplier for tires. Similarly, the final preference ratings for A_4, A_5, and A_6 are found to be 0.318, 0.340, and 0.342, respectively. This means that the three suppliers are roughly equal (although A_6 has the highest rating) in supplying electronics, a component that calls for a hybrid strategy. The final preference ratings for A_7, A_8, and A_9 are found to be 0.221, 0.463, and 0.317, respectively. This means that A_8 is the most preferred, agile-oriented supplier for peripherals. Note that this analysis result does match the characteristic assumption of the suppliers.

Although AHP is a popular approach for MCDM and has been used for supplier selection by many researchers, there is a debate regarding its mathematical rigor. Belton and Gear (1984) argued that AHP lacked a firm theoretical basis and is not an axiomatic approach like MAUT, hence it was a flawed theory in analyzing decisions. Satty (1986) then provided theorems to prove that AHP laid on an axiomatic theory and that the 1-9 preferential scale was an adept method to elicit preferences for the AHP method even though it was an open research topic with much to be explored. Dyer (1990) questioned the validity of Satty's axioms. Satty (1990), together with Harker and Vargas (1990), defended their standpoints that the axioms of AHP are different from that of traditional utility theory, and they are valid.

The most disputed issue in AHP is rank reversal. Three major arguing points are summarized as follows:

- Addition/deletion of alternatives—Belton and Gear (1984) criticized that any addition of alternatives caused a rank reversal in AHP. They illustrated this phenomenon with a numerical example where the rankings of alternatives based on pair-wise comparison matrices reversed in AHP if new alternatives were added. The reason for this preference reversal is that "the procedure is consistent with a belief that the relative importance of criteria is proportional to the arithmetic mean value of the options on each criterion." They then proposed that "a reasonable procedure is one which is consistent with a definition of weight as the value of a unit of scale on which the criterion is measured. The scale is determined by the nature of the option ranked most highly on it." Harker and Vargas (1987) and Satty (1986) defended the attack by indicating that "the rank reversal was because Belton and Gear applied MAUT weights on the AHP derived eigen vectors to derive the rankings, which is an incorrect method. If the AHP method of deriving the ranked preferences were followed, then the rank reversal would not have occurred. The weights that are considered to be equal in MAUT have to be transformed in AHP to preserve not only the preference but also the ratio among the values." Satty (1990) further pointed out that "with the absolute measurement of the AHP, there can never be reversal in the rank of the alternatives by adding or deleting other alternatives."

- Near copies and exact copies of alternatives. Belton and Gear (1984) also discussed the addition of exact copies of alternatives to cause rank reversal. Harker and Vargas (1987) criticized that Belton and Gear's counterexample is vacuous since Axiom 4 developed by Satty (1986) explicitly excludes exact copies from consideration. Dyer (1990) extended Belton and Gear's example to near copies of alternatives to question the rigor of AHP. Satty (1990) defended that "the presence of copies or near copies in absolute measurement in the AHP has no effect whatsoever on the rank of alternatives."

- Multiple choice versus single choice. Dyer (1990) pointed out that a static set of AHP weighting can lead to arbitrary rankings when multiple alternatives are selected at one time. For example, suppose there are three suppliers A, B, and C, in order of their AHP weighting preference. Now if A were selected first, it is possible that the AHP weightings of B and C might change if A were no longer included in the set of paired comparisons. Hence, the AHP weightings, according to Dyer, become arbitrary.

Satty (1990) and other researchers (Harker and Vargas 1990) disagreed with Dyer's opinion. Satty argued that "MAUT is a normative process, while AHP is a descriptive theory that encompasses procedures leading to outcomes as would be ranked by a normative theory." He pointed out that Dyer built certain expectation about the AHP, because he assumed that there is a unique way to deal with decision problems, more or less along the traditional lines of utility theory largely reflected in his own work. Satty (1990) also indicated two flaws in Dyer's logic. The first one is to do with change in criteria weights and rank reversal, and the second one is about the axioms and outcomes of AHP and MAUT. Harker and Vargas (1990) argued that the axioms of AHP provided by Satty (1986) are different from that of traditional utility theory, and they indicated the reason of rank reversal is because those alternatives depend on what alternatives are considered; hence, adding or deleting alternatives can lead to change in the rank. They also disagreed with the method proposed by Dyer and concluded that "contrary to utility theory, no mathematically unjustified rank reversal occurs in the AHP."

Recently, researchers have been working on the rank reversal dispute in much depth and have made some progress. Barzilai (1998) provided a mathematical framework for the decomposition of value functions into linear and nonlinear tree structures. One application of this framework is the analysis of the AHP. He argued that the AHP generates nonequivalent value functions and rankings from equivalent decompositions. Aguaron and Moreno-Jimenez (2000) provided a sensitivity analysis of the judgments used in the AHP in relation to the rank reversal produced in two different situations: the selection of the best alternative, and the ranking of all the alternatives. Millet and Satty (2000) indicated that AHP now supports two types of synthesis procedures—one that preserves rank (ideal mode) and one that does not (distributive mode). They proposed guidelines for the selection of one synthesis mode over the other, and also proposed several modifications to software implementations of the AHP to better support both synthesis modes and to reduce user stress over the choice. Hurley (2001) proposed a simple approach to sensitivity analysis within the AHP, which preserves the rank order of the objects. He indicated that if the weights of an AHP pair-wise comparison matrix can be varied in a way that preserves the rank order of the objects, and at the same time, this perturbation does not result in the best alternative changing, then the decision maker is typically much more confident about what the AHP recommends.

Dyer (1990) raised a minor objection to the Satty's 1-9 scale by illustrating an example, while Harker and Vargas (1990) argued that the use of 1-9 scale did not affect the theory of AHP by pointing out the misconception procedure of the example provided by Dyer. They stated that "empirical evidence suggests that it appears to be an appropriate scale to capture decision maker's preferences, …, the AHP is fairly robust even if 'errors' are made."

Wang et al. (2005) justified the use of AHP in supplier selection. AHP's hierarchic structure reflects the natural tendency of human mind to sort elements of a system into different levels and to group similar elements in each level, which can facilitate decision maker's easy understanding from a human factor point of view. A study conducted by Lehner and Zirk (1987) shows that when a human being and an intelligent machine cooperate to solve problems, but where each employs different problem-solving procedures, the user must have an accurate model of how that machine operates. This is because when people deal with complex, interactive systems, they usually build up their own conceptual mental model of the system. The model guides their actions and helps them interpret the system's behavior. Such a model, when appropriate, can be very helpful or even necessary for dealing successfully with the system. However, if inappropriate or inadequate, it can lead to serious misconceptions or errors (Young 1981). Therefore, it is very important for decision makers to be able to understand the decision-making model structure, while AHP just provides such a single, easily understood, and flexible model structure.

The supplier selection decision-making process is a rather complex problem, which involves both qualitative and quantitative factors. Decision makers play a very important role in this process. AHP allows the decision makers to reflect their "conceptual mental model" of the process. The aim of AHP is to facilitate the decision makers to understand the model structure and to aid the decision-making process, rather than to mathematically optimize the process. In addition, with regard to performance assessment and comparison, rank reversal is not particularly critical, since the following facts are observed:

- The set of criteria and subcriteria to be compared, which are supply chain performance metrics, does not change. Therefore, multiple choices would not cause rank reversal problem.
- Different alternatives are considered, so exact copies or near copies of alternatives do not exist. Therefore, no rank reversal will occur due to exact or near copies of alternatives.
- Absolute measurements (i.e., the ranking of the criteria is initially independent of the particular alternatives in the decision problem being considered) are used, so the addition and/or deletion of alternatives will not cause any rank reversal.

6.3.2 Multiattribute Utility Theory Approach

MAUT is a well-established and mathematically sound method for MCDM. The notion of utility originated from the domain of economics. In the course of the development of economics, it has been found that the key to this vast subject is the analysis of the behavior of the individuals constituting the economic community. A fundamental assumption about the motives of the individual in economic behaviors is that the consumer desires to obtain a maximum of utility or satisfaction and the entrepreneur a maximum of profit; the individual who behaves accordingly is said to be "rational." Concerned about how to mathematically model the question of rational behavior, Von Neumann and Morgenstern (1947) first established the theory of utility. They see the notion of utility as an immediate sensation of individual's subjective preference or satisfactory level toward alternatives; therefore, in a way similar to the quantification of heat and light in physics, a numerical measurement should be able to be developed to evaluate utility. This concept, which roots in the desires to employ a mathematical method in economics, gives one of the basic economic problems, that is, "to find an exact description of the endeavor of the individual to obtain a maximum of utility," a mathematical solution.

Quantification of utility is implemented by the combination of two "natural" concepts in the domain of utility: (1) one alternative is preferable to another, which implies a "greater" operator, and (2) one alternative is equally desirable to a combination of two other alternatives associated with probability of happening, α and $1 - \alpha$, respectively, which implies an "equal" operator.

The first concept is a common experience to every human being. The second one will also appear natural by considering the following gambling scenario. Hypothetically, one is given two choices, a $5 bill or a ticket to enter a lottery which will result in equal chance of winning $10 or nothing. Many people would feel these two alternatives equally desirable, which is the case of the second concept. Many others might prefer $5 cash, and still others might be more tempted to a 50% chance of earning $10. The attitudes toward uncertainties represented by these three groups are termed as risk neutral, risk averse, and risk prone, respectively.

The following example illustrates how these two concepts are exploited in the quantification. There are three alternatives, A, B, and C, with the preference structure such that C is preferred to both A and B, and B is preferred to A, and with a subjective assertion that B is equally desirable with the combined event consisting of a chance of probability $1 - \alpha$ for A and a chance of probability α for C. A function $\mu(x)$ is then postulated to mathematically describe this situation, and it should have the following properties:

$$\mu(C) > \mu(B) > \mu(A) \tag{6.1}$$

$$\mu(B) = (1-\alpha)\mu(A) + \alpha\mu(C) \tag{6.2}$$

Equation 6.2 can be rewritten as

$$\alpha = \frac{[\mu(B) - \mu(A)]}{[\mu(C) - \mu(A)]} \tag{6.3}$$

Therefore, α could be used as a numerical estimate for the ratio of the preference of B over A to that of C over A. Thus, if $\mu(A)$ and $\mu(C)$ are arbitrarily set as 0 and 1, respectively, then $\mu(B)$ will have a utility value of α.

These concepts are the basis of MAUT, which provides a mathematically rigorous method to compare alternatives. Here, we use an example to show how MAUT can be used for supplier selection.

Ultra Inc. is a personal computer (PC) manufacturer with a business model based on modularization and postponement of product differentiation. It purchases most of the components from suppliers and assembles the final product prior to delivering to the customer. The company has recently developed a new model Ultra Pro based on a high-performance memory chip. Initial forecast shows that introduction of this new model will generate heavy demand. Since the product is innovative and in the introduction and growth stage of its life cycle, management has decided to emphasize flexibility and responsiveness as the demand may fluctuate significantly. In addition, management has decided to select from three global suppliers (S_1, S_2, and S_3) with whom the company has a strategic partnership to support the production of this new PC model because of its importance to the company's

TABLE 6.11

Selected Metrics and Supplier Performance Data

Category	Metrics	Performance			Utility			Scale Factor
		S_1	S_2	S_3	S_1	S_2	S_3	
Reliability	% Orders received defect free	95%	97%	87%	0.91	0.93	0.82	0.15
	Fill rate	97%	95%	91%	0.92	0.89	0.78	0.10
	In process failure rate	93%	97%	95%	0.87	0.94	0.91	0.05
Responsiveness	Published delivery cycle time	28	36	12	0.89	0.75	0.92	0.25
Flexibility	Cost of expediting delivery process	15	22	15	0.92	0.87	0.92	0.25
Cost and financial	Warranty cost	12	8	9	0.81	0.96	0.94	0.05
	Tariffs and custom duties	20	32	23	0.94	0.88	0.93	0.10
	Inventory carrying cost	40	33	33	0.90	0.93	0.93	0.05

Source: *Int. J. Prod. Econ.*, 105(2), Huang, S.H., and Keskar, H., Comprehensive and configurable metrics for supplier selection, 510–523, Copyright 2007, with permission from Elsevier.

profitability. All three suppliers have similar assets and infrastructure. In addition, all of them have safe and environmental-friendly production facility. Therefore, the management at Ultra decided to consider only reliability, responsiveness, flexibility, and cost and financial metrics. A total of eight metrics were selected as shown in Table 6.11.

To use MAUT for decision making, utility function for each attribute (performance metric under consideration) needs to be established first to capture management's preference. There are various techniques for determining utility functions (Keeney and Raiffa 1976). The most common one is the mid-value splitting technique that involves an analyst (can be a computer program) interviewing the decision maker. Take published delivery cycle time as an example, the analyst will first ask the decision maker what is considered a perfect cycle time and what is absolutely unacceptable. Suppose the decision maker answers 4 h and 96 h, respectively, then a utility of 1 is assigned to 4 h and a utility of 0 is assigned to 96 h. The analyst then asks, "Suppose a supplier improves from a 96 h delivery cycle time to $h_{0.5}$ h and another supplier improves from $h_{0.5}$ h to 4 h, what would be the value of $h_{0.5}$ so that you think the two suppliers have equal amount of improvement?" Suppose the answer is 54 h, then a utility of 0.5 is assigned to 54 h. The analyst will then ask a similar question—"Suppose a supplier improves from a 54 h delivery cycle time to $h_{0.75}$ h and another supplier improves from $h_{0.75}$ h to 4 h, what would be the value of $h_{0.75}$ so that you think the two suppliers have equal amount of improvement?" If the decision maker answers 36 h, then 36 h is assigned a utility of 0.75. Once a few data points (5–7) are obtained, the analyst will fit these data points into a curve and ask the decision maker to evaluate and fine-tune the curve to obtain a desired utility function. When the utility function is established, it can be used to obtain a utility value for a specific supplier based on its performance.

For each metric, this mid-value splitting technique is used to establish a utility function. The utility values obtained using these utility functions for all three suppliers are shown in Table 6.11. Also shown in Table 6.11 is the scale factor for each metric. The scale factor is obtained through an inter-view-based process similar to that used to establish a utility function. It is a rather elaborated process and may be confusing at times. Interested readers can refer to Keeney and Raiffa (1976) for more details. With the scale factor and all the utility values determined, the evaluation of suppliers in this case is nothing but a weight sum problem. We found that the overall utility values for suppliers S_1, S_2, and S_3 are 0.903, 0.863, and 0.893, respectively. Thus, S_1 is the most preferred supplier, the preference score of S_3 is not far behind, whereas S_2 is the least preferred supplier.

6.4 Contracts to Increase Supply Chain Profitability

Once the potential suppliers are ranked, a company will proceed to negotiate sales contracts with the highest ranked suppliers. Sales contracts should be designed to increase supply chain profitability. However, contract negotiation involves two parties, each aiming to optimize its own profit. The result is often lower profits than what would be achieved if the two parties coordinate their actions with a common objective of maximizing supply chain profitability.

Consider the case of Ultra Inc. The company priced the new PC model, Ultra Pro, at $1500. At this price, the demand is estimated to be normally distributed with a mean of 300,000 units and a standard deviation of 30,000 units. There is a possibility that Ultra Pro may not be successful. In that case Ultra will discontinue its production and any unsold Ultra Pro will be heavily discounted. The company estimated that at a discounted price of $500 per unit all remaining Ultra Pro can be sold to a wholesaler. The production cost of Ultra Pro is $400 per unit. The company has decided to purchase components from supplier S_1, whose cost for producing the components for one unit of Ultra Pro is $500. Supplier S_1 quoted a sales price of $800 to Ultra. Note that for the entire supply chain, the cost for producing one unit of Ultra Pro is $900. The optimal cycle service level is $CSL^* = \dfrac{p-c}{p-s} = \dfrac{1500-900}{1500-500} = 0.6$. The optimal production quantity is $Q^* = F^{-1}(0.6, 300000, 30000) = 307,600$ units. The expected gross profit is

$E(Q^*)=(1500-500)\times 300000\times F(307600,300000,30000)-(1500-500)\times 30000^2\times$
$f(307600,300000,30000)-(900-500)\times 307600\times F(307600,300000,30000)+$
$(1500-900)\times 307600\times[1-F(307600,300000,30000)]=\$168,409,724.$

However, because supplier S_1 quoted a unit component price of \$800 to Ultra. Ultra's cost of producing one unit of Ultra Pro is \$1200. Its optimal cycle service level is $CSL^*=\dfrac{p-c}{p-s}=\dfrac{1500-1200}{1500-500}=0.3$. Its optimal production quantity is $Q^*=F^{-1}(0.3,300000,30000)=284,268$ units, which is quite conservative. As a result, its expected gross profit is \$79,569,222. Because supplier S_1 sold 284,268 units of components, it made a gross profit of $284,268\times\$300=\$85,280,400$. The result is a total supply chain gross profit of \$164,849,622, which is less than the optimal supply chain gross profit.

A *buyback contract* between Ultra and the supplier can alleviate this problem. In the contract, supplier S_1 agrees to buy all unsold Ultra Pro at a unit price of \$550. It then sells them to the wholesaler for \$500. This means that the supplier will incur a cost of \$50 for each unit of unsold Ultra Pro. The effect of this buyback clause to Ultra is that it raises the salvage value to \$550. As a result, Ultra's optimal CSL increases to 0.316. Its optimal production quantity increases to 285,615 units. Its expected gross profit is \$79,864,920, higher than that without a buyback clause. The expected overstock is $(285615-300000)\times F(285615,300000,30000)+30000^2\times f(285615,300000,30000)=6,126$ units. For the supplier, it made a gross profit of \$85,684,500 by selling 285,615 units of component, but it is expected to incur a cost of $6126\times\$50=\$306,300$ because of the buyback clause. This results in an expected gross profit of \$85,378,200, which is also higher than that without the buyback clause.

Ultra and the supplier can also use a *revenue sharing* contract. In the contract, the supplier agrees to sell the components at cost (\$500 per unit) to Ultra; and Ultra agrees to share 20% of the revenue with the supplier. In other words, for each unit of Ultra Pro sold, the supplier will receive an additional \$300 ($\$1500\times20\%$). In this case, the sales revenue that Ultra received is \$1200 per unit and its cost is \$900 per unit. The salvage value remains at \$500. As a result, Ultra's optimal CSL increases to 0.429. Its optimal production quantity increases to 294,600 units. The expected overstock is 9462 units. Ultra is expected to sell $294,600-9,462=285,138$ units. Its expected gross profit is $285,138\times(1-20\%)\times\$1,500+9,462\times\$500-294,600\times\$900=\$81,756,600$. The supplier has an expected gross profit of $285,138\times\$300=\$85,541,510$.

One can see from the aforementioned example that the revenue-sharing contract produces higher total supply chain profit than that with the buyback contract. In addition, the unsold products do not need to be returned to the supplier. However, such a contract requires an information infrastructure that allows the supplier to monitor the sales, which could be expensive to build.

6.5 State of the Art in Supplier Selection

The subject of supplier selection has been studied extensively in recent years. However, unlike other more mature subjects such as forecasting, aggregate planning, inventory management, and distribution network design and transportation, no consensus has been reached on specific analytical methods for supplier selection. In addition to AHP and MAUT, other MCDM methods have also been used to select suppliers, including data envelopment analysis (DEA), mathematical programming, case-based reasoning, analytic network process, fuzzy set theory, and genetic algorithms. Furthermore, a number of integrated approaches that combine different MCDM methods have also been used. Ho et al. (2010) conducted a comprehensive survey of these methods. The most popular stand-alone MCDM method used for supplier selection was DEA. DEA was originally developed to evaluate the efficiency of a number of producers, or decision-making units (DMUs), with common inputs and outputs (Charnes et al. 1978). It is a mathematical programming model applied to observational data that provides a new way of obtaining empirical estimates of extreme relations. Instead of trying to fit a regression plane through the center of the data, DEA "floats" a piecewise linear surface to rest on top of the observations. Because of this unique perspective, DEA is able to uncover relationships that remain hidden for other methodologies (Seiford and Thrall 1990).

The basic assumption of DEA is that if a DMU is capable of producing y units of output given x units of input, then all other DMUs should be able to do the same if they were to operate with the same level of efficiency. Therefore, all the DMUs can be combined to form a composite "virtual" DMU with composite inputs and outputs. Finding the "best" virtual DMU can be formulated as a linear programming problem. Compared to the original DMU, if the "best" virtual DMU can produce more output with the same input (or produce the same output with less input), then it is more efficient. In other words, the original DMU is inefficient. This idea is used to determine the efficiency of a supplier. Note that originally DEA was used to measure the relative efficiencies of homogeneous DMUs based on numerical data alone. In supplier selection, researchers have modified the DEA method to handle both quantitative and qualitative criteria. Considerations have also been given to stochastic performance measures and imprecise data.

The popularity of DEA is a result of its sound mathematical foundation and its robustness in real-world application. However, Ho et al. (2010) identified three limitations of DEA for supplier selection. First, confusion exists among practitioners with respect to the input and output criteria. For example, some researchers consider price/cost as an output criterion, whereas others consider it as an input criterion. Second, the assignment of ratings to qualitative criteria is subjective, which at times leads to inconsistent results. Third, DEA was originally developed to measure the relative output/input efficiencies of homogenous DMUs. It is not clear whether the most efficient supplier, as determined through DEA, is also the most effective supplier.

AHP is also a popular approach and is often combined with other methods for supplier selection. AHP is conceptually simpler, flexible, and easier to use. It has a formal consistency checking procedure that acts as a feedback mechanism for decision makers to review and revise their subjective judgment. The outcome of AHP, the same as the outcome of MAUT, is relative weights of the suppliers. Resource constraints such as supplier capacity are not considered during the process of deriving the weights. Therefore, AHP weights are often incorporated into a goal programming model to determine the optimal set of suppliers and the ordering quantity from each supplier, subject to resource constraints. A potential drawback of AHP, pointed out by Ho et al. (2010), is that the pair-wise comparison process may be time consuming, especially when inconsistency is detected that requires further revision of expert judgment.

Ho et al. (2010) also identified a list of the most popular criteria used to evaluate supplier performance, including quality, delivery, price/cost, manufacturing capability, service, management, technology, research and development, finance, flexibility, reputation, relationship, risk, and safety and environment. Their analysis indicated that the traditional single criterion approach based on lowest cost for supplier selection is no longer sufficient in current supply chain management practice. It was recommended that when defining criteria for supplier selection, a company should consider its business objectives and the requirements of company stakeholders. Specifically, quality function deployment was proposed as a tool to map business objectives and company stakeholder requirements into a set of supplier performance evaluation criteria.

References

Aguaron, J. and J. M. Moreno-Jimenez. 2000. Local stability intervals in the analytic hierarchy process. *European Journal of Operational Research* **125**: 113–132.

Barzilai, J. 1998. On the decomposition of value functions. *Journal Operations Research Letters* **22**: 159–170.

Belton, V. and T. Gear. 1984. On the shortcoming of Saaty's method of analytic hierarchies. *Omega* **11**: 228–230.

Charnes, A., W. W. Cooper, and E. Rhodes. 1978. Measuring the efficiency of decision making units. *European Journal of Operational Research* **2**: 429–444.

Chopra, S. and P. Meindl. 2010. *Supply Chain Management: Strategy, Planning, & Operation*, 4th edn. Upper Saddle River, NJ: Pearson Prentice Hall.

Dyer, J. S. 1990. Remarks on the analytic hierarchy process. *Management Science* **36**: 249–258.

Harker, P. T. and L. G. Vargas. 1987. The theory of ratio scale estimation: Satty's analytic hierarchy process. *Management Science* **33**: 1383–1403.

Harker, P. T. and L. G. Vargas. 1990. Reply to Remarks on the analytic hierarchy process by J. S. Dyer. *Management Science* **36**: 269–273.

Ho, W., X. Xu, and P. K. Dey. 2010. Multi-criteria decision making approaches for supplier evaluation and selection: A literature review. *European Journal of Operational Research* **202**: 16–24.

Huang, S. H. and H. Keskar. 2007. Comprehensive and configurable metrics for supplier selection. *International Journal of Production Economics* **105**(2): 510–523.

Hurley, W. J. 2001. Analytic hierarchy process: A note on an approach to sensitivity which preserves rank order. *Computers and Operations Research* **28**: 185–188.

Keeney, R. L. and H. Raiffa. 1976. *Decisions with Multiple Objectives: Preferences and Value Tradeoffs.* New York: Wiley.

Krajewski, L. J. and L. P. Ritzman. 2001. *Operations Management: Strategy and Analysis*, 6th edn. Reading, MA: Addison Wesley.

Lehner, P. E. and D. A. Zirk. 1987. Cognitive factors in user/expert system interaction. *Human Factors* **29**: 97–109.

Millet, I. and T. L. Satty. 2000. On the relativity of relative measures – Accommodating both rank preservation and rank reversals in the AHP. *European Journal of Operational Research* **121**: 205–212.

Saaty, T. L. 1980. *The Analytic Hierarchy Process.* New York: McGraw-Hill.

Satty, T. L. 1986. Axiomatic foundation of the analytic hierarchy process. *Management Science* **32**: 841–855.

Satty, T. L. 1990. An exposition of the AHP in reply to the paper Remarks on the analytic hierarchy process. *Management Science* **36**: 259–268.

Seiford, L. M. and R. M. Thrall. 1990. Recent developments in DEA: The mathematical programming approach to frontier analysis. *Journal of Econometrics* **46**: 7–38.

Tummala, V. M. R. and C. L. Tang. 1996. Strategic quality management, Malcolm Baldrige and European quality awards and ISO 9000 certification: Core concepts and comparative analysis. *International Journal of Quality* **13**: 8–38.

Von Neumann, J. and O. Morgenstern. 1947. *Theory of Games and Economic Behavior*, 2nd edn. Princeton, NJ: Princeton University Press.

Wang, G., S. H. Huang, and J. P. Dismukes. 2004, Product driven supply chain selection using integrated multi-criteria decision making methodology. *International Journal of Production Economics* **91**(1): 1–15.

Wang, G., S. H. Huang, and J. P. Dismukes. 2005. Manufacturing supply chain design and evaluation. *International Journal of Advanced Manufacturing Technology* **25**: 93–100.

Young, R. M. 1981. The machine inside the machine: Users' models of pocket calculators. *International Journal of Man-Machine Study* **15**: 51–85.

7

Supply Chain Simulation Game

7.1 Overview

This chapter provides a simulation game for students to practice the supply chain management skills they acquired. The simulation game involves eight companies and requires eight student participants. Each student will operate one of the eight companies and interact with others to establish a supply chain. The operation of the supply chain is then simulated (with a 1-year time frame) to allow students to observe the impact of their decisions on company and supply chain profitability. The simulation game is described as follows.

Two laptop manufacturers, Bell Inc. and Ultra Corp., dominate the market. The laptop manufacturers outsource two components, DVD drive and battery. There are two DVD drive suppliers, Alpha and Beta, and two battery suppliers, Longey and Everlast. The laptops are sold through two retailers, BVE Store and Pop Electronics. Purchase orders (for laptops and components) are placed at the beginning of each month. All companies use TransMono for transportation of products. TransMono delivers products overnight once an order is received. It charges a fixed $5000 service fee for each delivery, plus $1 for each DVD drive or battery and $2 for each laptop delivered. A delivery is one that has a single destination with one or more pick-up locations. For example, if Bell Inc. orders 1000 DVD drives from Beta and 1000 batteries from Longey and have TransMono deliver them at the same time, it will incur a shipping cost of $7000. This basic information of the supply chain is shown in Table 7.1.

The retail prices for Bell and Ultra laptops are $1000 per unit and $1700 per unit, respectively. Demand during stockout is lost. The inventory holding costs at the retailers are $12 and $15 per unit per month for Bell and Ultra laptops, respectively. Operating expense (fixed cost) is $1.5 million per month at both retailers. The initial laptop inventories are 0 at both retailers. This basic information of the retailers is shown in Table 7.2. The monthly laptop sales of the retailers for the past 5 years are shown in Tables 7.3 through 7.6. Each retailer needs to negotiate sales contracts with and provide monthly order quantities to both laptop manufacturers.

TABLE 7.1

Basic Supply Chain Information

	BVE Store
Retailer	Pop Electronics
Laptop manufacturer	Bell Inc.
	Ultra Corp.
DVD drive supplier	Alpha
	Beta
Battery supplier	Longey
	Everlast
Purchase order	Placed at the beginning of each month
Delivery lead time	Overnight
Transportation cost	$5000 for each delivery to a single customer, plus $2 for each laptop and $1 for each DVD drive or battery delivered

TABLE 7.2

Basic Retailer Information

	Bell Laptop	Ultra Laptop
Retail price	$1,000/unit	$1,700/unit
Inventory holding cost	$12/unit/month	$15/unit/month
Initial inventory	0	0
Operating expense	$1.5 million/month	
Demand during stockout	Lost	

TABLE 7.3

Sales of Bell Laptops at BVE Store

Month	Year				
	2008	2009	2010	2011	2012
1	904	902	1066	990	999
2	940	947	1003	1009	977
3	1074	865	1021	942	1004
4	959	975	971	1053	1013
5	1095	938	995	940	989
6	1230	1155	1212	1164	1152
7	1232	1172	1217	1211	1210
8	1008	950	983	948	973
9	987	1068	1022	1006	889
10	1041	1013	1029	980	914
11	1344	1400	1352	1403	1353
12	1419	1449	1425	1373	1305

TABLE 7.4

Sales of Ultra Laptops at BVE Store

Month	Year				
	2008	2009	2010	2011	2012
1	3173	2869	3694	3191	3240
2	2929	3453	2150	2247	3076
3	3071	2925	2648	3015	3778
4	2967	3397	2819	2705	2216
5	2437	2703	3136	3817	2984
6	2566	2883	2465	2267	2518
7	2951	2699	2473	3223	3249
8	3367	2549	3284	2498	2155
9	2997	3223	2571	3269	3410
10	3252	2640	3132	2697	3516
11	2685	2981	3008	3222	2888
12	2938	2987	2902	2725	3254

TABLE 7.5

Sales Figures of Bell Laptops at Pop Electronics

Month	Year				
	2008	2009	2010	2011	2012
1	4862	5224	4805	4953	5164
2	5231	4992	4702	5005	4915
3	5211	4484	4761	5522	4753
4	4780	4847	4781	5244	4864
5	4677	4789	5005	5073	5067
6	5520	5997	5963	6453	6264
7	5112	6486	5768	6310	5827
8	4842	5026	5327	5129	5688
9	5211	4594	4649	5409	5546
10	5147	4836	4813	5086	5101
11	6443	6846	7056	7046	7015
12	7049	6744	7177	6632	7049

The material costs (excluding DVD drive and battery) for producing a Bell laptop and an Ultra laptop are $400 and $600, respectively. The labor hour required for producing a laptop is 1 per unit. A maximum of 160 regular time hours and 40 overtime hours are available each month. Workers are paid a monthly salary of $1500 each no matter how many regular time hours they work. For overtime work, each worker is paid $14 an hour. The costs of hiring and layoff a worker is $1000 and $2000, respectively. The inventory holding cost for a laptop is $10 per month. The inventory holding cost for a DVD drive

TABLE 7.6

Sales of Ultra Laptops at Pop Electronics

Month	Year				
	2008	2009	2010	2011	2012
1	941	1013	1040	1131	803
2	1132	1161	912	893	1209
3	987	994	957	1133	1066
4	920	864	1006	864	707
5	675	821	1048	821	826
6	1049	1133	941	944	1249
7	981	828	972	1054	1055
8	1141	871	933	1034	1216
9	797	1149	998	1041	750
10	1148	944	1106	809	1129
11	1051	785	1073	1079	1129
12	864	960	891	1004	1161

or a battery is $4 per month. Operating expense is $300,000 per month at both laptop manufacturers. Each laptop manufacturer has a manufacturing facility that can accommodate a maximum of 80 workers. The initial workforce is 0 at both laptop manufacturers. The initial laptop inventory is 8000 units at Bell Inc. and 5000 units at Ultra Corp. The initial component (DVD drive and battery) inventories at both laptop manufacturers are 0. This basic information of the laptop manufacturers is shown in Table 7.7. In addition to

TABLE 7.7

Basic Laptop Manufacturer Information

	Bell Inc.	Ultra Corp.
Material cost	$400/unit	$600/unit
Initial inventory	8000	5000
Labor hour required	1 worker hour/unit	
Maximum regular time	160 hours/worker	
Maximum overtime	40 h/worker	
Regular time pay	$1500/month/worker	
Overtime pay	$14/h/worker	
Hiring cost	$1,000/worker	
Layoff cost	$2,000/worker	
Laptop holding cost	$10/unit/month	
DVD drive holding cost	$4/unit/month	
Battery holding cost	$4/unit/month	
Operating expense	$300,000/month	
Maximum workforce	80 workers	
Initial workforce	0	

TABLE 7.8

Basic Component Supplier Information

Material cost	$40/unit
Labor hour required	1 worker hour/unit
Maximum regular time	160 h/worker
Maximum overtime	40 h/worker
Regular time pay	$1,440/month/worker
Overtime pay	$12/h/worker
Hiring cost	$500/worker
Layoff cost	$1,000/worker
Inventory holding cost	$3/unit/month
Operating expense	$100,000/month
Maximum workforce	40 workers
Initial workforce	0
Initial inventory	8,000

negotiating sales contracts with the retailers, each laptop manufacturer needs to negotiate sales contracts with and provide monthly order quantities to the DVD and battery suppliers. Each manufacturer also needs to develop an aggregate plan.

The material cost for a DVD drive or a battery is $40. The labor hour required for producing a DVD drive or a battery is 1 per unit. A maximum of 160 regular time hours and 40 overtime hours are available each month. Workers are paid a monthly salary of $1440 each no matter how many regular time hours they work. For overtime work, each worker is paid $12 an hour. The cost of hiring and laying off a worker is $500 and $1000, respectively. The inventory holding cost for a DVD drive or a battery is $3 per month. Operating expense is $100,000 per month at both suppliers. Each supplier has a manufacturing facility that can accommodate a maximum of 40 workers. The initial workforce is 0 and the initial inventories are 8000 at all suppliers. This basic information of the component suppliers is shown in Table 7.8. In addition to negotiating sales contract with the laptop manufacturers, each supplier needs to develop an aggregate plan.

7.2 Sales Contract and Aggregate Plan

Each retailer needs to negotiate two sales contracts; one with Bell Inc. and the other with Ultra Corp. Each laptop manufacturer, in addition to the sales contracts with the retailers, needs to negotiate with at least one DVD supplier and at least one battery supplier to ensure that it has the required components for laptop production. A sales contract should contain the following information: (1) unit price using all-unit quantity discount with three price breaks; (2) which party (buyer or seller) pays for the shipping cost; (3) unit penalty for the portion of the demand that cannot be fulfilled by the seller; and (4) estimates of monthly demand for a year. Tables 7.9 through 7.13 show sample sales contract between BVE and Bell Inc., between BVE and Ultra Corp., between Pop Electronics and Bell, between Bell and Longey, and between Bell and Beta.

Laptop manufacturers and component suppliers need to develop aggregate plans. The aggregate plan includes the production level and number of workers for each month. The required overtime hours, if any, can be calculated based on the production level and the number of workers available. Note that each month's order, received at the beginning of the month, is satisfied using existing inventory. Therefore, the production level in a certain month is determined based on the estimated demand from the subsequent month. The production level in December is determined based on the goal of bringing the ending inventory back to the initial inventory at the beginning of January. An aggregate plan must be feasible, that is, the number of workers must be sufficient to meet the production level. Note that an aggregate plan with more workers than necessary is also a feasible one. A sample aggregate plan for Bell Inc. is shown in Figure 7.1, including the calculation of overtime hours and determination of its feasibility.

TABLE 7.9

Sample Sales Contract between BVE (Buyer) and Bell Inc. (Seller)

Quantity	Unit Price
0–1000	$700
1001–2000	$680
2001 and above	$675
Shipping cost to be paid by	Buyer
Unit penalty for any unmet demand	$50
Monthly Demand	
January	1200
February	1001
March	1001
April	1001
May	1001
June	1200
July	1200
August	1001
September	1001
October	1001
November	1300
December	1300

TABLE 7.10

Sample Sales Contract between BVE (Buyer) and Ultra Corp. (Seller)

Quantity	Unit Price
0–2000	$1100
2001–4000	$1000
4001 and above	$950
Shipping cost to be paid by	Buyer
Unit penalty for any unmet demand	$75
Monthly Demand	
January	4001
February	3000
March	3000
April	3000
May	3000
June	3000
July	3000
August	3000
September	3000
October	3000
November	3000
December	3000

TABLE 7.11

Sample Sales Contract between Pop Electronics (Buyer) and Bell Inc. (Seller)

Quantity	Unit Price
0–4000	$750
4001–6000	$700
6001 and above	$680
Shipping cost to be paid by	Seller
Unit penalty for any unmet demand	$100
Monthly Demand	
January	6001
February	5000
March	5000
April	5000
May	5000
June	6001
July	6001
August	5000
September	5000
October	5000
November	7000
December	7000

TABLE 7.12

Sample Sales Contract between Bell Inc.
(Buyer) and Longey (Seller)

Quantity	Unit Price
0–4000	$75
4001–8000	$70
8001 and above	$65
Shipping cost to be paid by	Seller
Unit penalty for any unmet demand	$15
Monthly Demand	
January	8010
February	4001
March	6000
April	6000
May	7200
June	7200
July	6000
August	6000
September	6000
October	8300
November	8300
December	8001

TABLE 7.13

Sample Sales Contract between Bell Inc.
(Buyer) and Beta (Seller)

Quantity	Unit Price
0–3000	$75
3001–6000	$72
6001 and above	$70
Shipping cost to be paid by	Seller
Unit penalty for any unmet demand	$10
Monthly Demand	
January	6001
February	6001
March	6001
April	6001
May	7201
June	7201
July	6001
August	6001
September	6001
October	8300
November	8300
December	7500

	A	B	C	D	E
1	Month	Predetermined Production	Workers	Overtime	Feasibility
2	January	6001	35	401	Yes
3	February	6001	35	401	Yes
4	March	6001	35	401	Yes
5	April	6001	35	401	Yes
6	May	7201	38	1121	Yes
7	June	7201	38	1121	Yes
8	July	6001	35	401	Yes
9	August	6001	35	401	Yes
10	September	6001	35	401	Yes
11	October	8300	42	1580	Yes
12	November	8300	42	1580	Yes
13	December	7500	42	780	Yes

Cell	Formula	Note
D2	=IF(B2>C2*160,B2-C2*160,0)	Drag down to D13
E2	=IF(C2*40>D2,"YES"/"NO")	Drag down to E13

FIGURE 7.1
A sample aggregate plan for Bell Inc.

7.3 Simulation and Profit Analysis

Once all the sales contracts are negotiated and all the aggregate plans are finalized, the supply chain operation can be simulated given "actual" demand at the retailers. A sample "actual" monthly demand at BVE is shown in Table 7.14. In general, a buyer may choose to change its monthly order quantity after observing actual demand. However, to simplify the simulation, we only allow the buyer to choose one of the following two strategies: (1) order based on the estimated monthly demand given in the sales contract, and (2) specify a desired level of safety inventory and adjust the monthly order based on inventory on hand and estimated demand. Suppose BVE orders Ultra laptops based on the estimated monthly demand given in the sales contract but aims to maintain a safety inventory of 200 Bell laptops; its actual monthly order, given the actual demand shown in Table 7.14, can be determined as shown in Figure 7.2.

A laptop manufacturer or component supplier can also choose one of the following strategies with respect to the actual monthly production level: (1) follow the predetermined production level in the aggregate plan, and (2) specify a desired level of safety inventory and adjust the monthly production level based on inventory on hand and the predetermined production level. Note that the predetermined workforce size cannot be changed, which might constrain the actual production level. In addition, component availability at the laptop

TABLE 7.14

"Actual" Monthly Demand at BVE

Month	Bell Laptop	Ultra Laptop
January	1120	2955
February	1059	3611
March	1072	2278
April	1119	2644
May	1055	3347
June	1132	2305
July	1186	2575
August	991	3179
September	1095	3550
October	1025	3603
November	1539	3258
December	1348	2754

	A	B	C	D	E	F	G	H	I
1		Estimated Demand		Actual Demand		Inventory		Acutal Order	
2	Month	Bell Laptop	Ultra Laptop	Bell Laptop	Ultra Laptop	Bell Laptop	Ultra Laptop	Bell Laptop	Ultra Laptop
3						0	0		
4	January	1200	4001	1120	2955	80	1046	1200	4001
5	February	1001	3000	1059	3611	142	435	1121	3000
6	March	1001	3000	1072	2278	129	1157	1059	3000
7	April	1001	3000	1119	2644	82	1513	1072	3000
8	May	1001	3000	1055	3347	146	1166	1119	3000
9	June	1200	3000	1132	2305	268	1861	1254	3000
10	July	1200	3000	1186	2575	214	2286	1132	3000
11	August	1001	3000	991	3179	210	2107	987	3000
12	September	1001	3000	1095	3550	106	1557	991	3000
13	October	1001	3000	1025	3603	176	954	1095	3000
14	November	1300	3000	1539	3258	0	696	1324	3000
15	December	1300	3000	1348	2754	152	942	1500	3000
16		Bell Laptop	Ultra Laptop						
17	Safety Inventory	200							

Cell	Formula	Note
F4	=IF(F3+H4-D4>0,F3+H4-D4,0)	Drag right to G4 then drag down to G15
H4	=B4	Drag right to I4
H5	=IF(B$17= "",B5,B5+B$17-F4)	Drag right to I5 then drag down to I15

FIGURE 7.2
Calculation of actual monthly order at BVE given an ordering strategy.

manufacturers could also constrain the actual production level. For simplicity sake, the laptop manufacturers always order batteries and DVD drives based on the estimated monthly demands specified in the sales contracts.

We now use the sample aggregate plan for Bell Inc. shown in Figure 7.1 as an example. Note that Bell orders batteries and DVD drives based on the estimated monthly demands specified in the sales contract with Longey (Table 7.12) and Beta (Table 7.13), respectively. Assume the demands are fully satisfied by the two suppliers. The actual monthly orders from BVE are shown in Column H in Figure 7.2, whereas the actual monthly orders from Pop Electronics are the same as the estimated demands in the sales contract (Table 7.11). Bell Inc. aims to maintain a safety inventory of 500 units at the beginning of each month after orders are delivered. The calculation of the actual monthly production level, actual month sales to BVE and Pop Electronics, and monthly beginning inventory levels (before product delivery) is shown in Figure 7.3. Note that in the case that the laptop inventory on hand is not enough to satisfy both retailers, priority is given to Pop Electronics. This is because it is more profitable to sell to Pop Electronics. Based on the sales contracts, at any given month the unit sales price of Bell laptops to Pop Electronics is $700 and the penalty is $100 per unit, whereas the unit sales price of Bell laptops to BVE is either $700 (August and September) or $680 (the rest of the months) and the penalty is only $50 per unit (sales price to Pop Electronics – penalty to BVE > sales price to BVE – penalty to Pop Electronics).

With the information shown in Figure 7.3, we can calculate the profit for Bell Inc. as shown in Figure 7.4. The inventory holding cost is calculated by averaging the beginning inventory and the ending inventory in each month. The beginning inventory for laptops excludes those shipped to the retailers, whereas the beginning inventories for components include those purchased from the suppliers. The value of the remaining inventory at the end of December (Cell D29) is calculated by assuming the lowest selling price for laptops (based on the sales contracts) and using the most recent purchase price for batteries and DVD drives. For the component suppliers, the actual monthly production, sales to laptop manufacturers, and profit can be calculated in a similar way.

Month	Predetermined Production	Workers	Battery Purchased	DVD Drive Purchased	BVE Order	Pop Order	Actual Production	Sales to BVE	Sales to Pop	Laptop Inventory	Battery Inventory	DVD Drive Inventory
						Priority				8000	0	0
January	6001	35	8010	6001	1200	6001	5822	1200	6001	6621	2188	179
February	6001	35	4001	6001	1121	5000	6059	1121	5000	6559	130	121
March	6001	35	6000	6001	1059	5000	6072	1059	5000	6572	58	50
April	6001	35	6000	6001	1072	5000	6051	1072	5000	6551	7	0
May	7201	38	7200	7201	1119	5000	7201	1119	5000	7633	6	0
June	7201	38	7200	7201	1254	6001	7201	1254	6001	7579	5	0
July	6001	35	6000	6001	1132	6001	6001	1132	6001	6447	4	0
August	6001	35	6000	6001	987	5000	6001	987	5000	6461	3	0
September	6001	35	6000	6001	991	5000	6001	991	5000	6471	2	0
October	8300	42	8300	8300	1095	5000	8300	1095	5000	8676	2	0
November	8300	42	8300	8300	1324	7000	8300	1324	7000	8652	2	0
December	7500	42	8001	7500	1500	7000	7500	1500	7000	7652	503	0

Cell	Formula	Note
H3	=MIN(500+F4+G4−(K2−I3−J3),C3*(160+40),L2+D3,M2+E3)	Drag down to H13
H14	=MIN(8000−(K13−I14−J14),C14*(160+40),L13+D14,M13+E14)	
I3	=IF(K2−J3−F3>0,F3,K2−J3)	Drag down to I14
J3	=IF(K2>G3,G3,K2)	Drag down to J14
K3	=K2+H3−I3−J3	Drag down to K14
L3	=D3−H3+L2	Drag down to L14
M3	=E3−H3+M2	Drag down to M14

FIGURE 7.3
Calculation of actual production and sales for Bell Inc.

Month	Workers	Battery Purchased	DVD Drive Purchased	Actual Production	BVE Order	Pop Order	Sales to BVE	Sales to Pop	Laptop Inventory	Battery Inventory	DVD Drive Inventory	Overtime	Hiring	Layoff	
	0								8000	0	0				
January	35	8010	6001	5822	1200	6001	1200	6001	6621	2188	179	222	35	0	
February	35	4001	6001	6059	1121	5000	1121	5000	6559	130	121	459	0	0	
March	35	6000	6001	6072	1059	5000	1059	5000	6572	58	50	472	0	0	
April	35	6000	6001	6051	1072	5000	1072	5000	6551	7	0	451	0	0	
May	38	7200	7201	7201	1119	5000	1119	5000	7633	6	0	1121	3	0	
June	38	7200	7201	7201	1254	6001	1254	6001	7579	5	0	1121	0	0	
July	35	6000	6001	6001	1132	6001	1132	6001	6447	4	0	401	0	3	
August	35	6000	6001	6001	987	5000	987	5000	6461	3	0	401	0	0	
September	35	6000	6001	6001	991	5000	991	5000	6471	2	0	401	0	0	
October	42	8300	8300	8300	1095	5000	1095	5000	8676	2	0	1580	7	0	
November	42	8300	8300	8300	1324	7000	1324	7000	8652	2	0	1580	0	0	
December	42	8001	7500	7500	1500	7000	1500	7000	7652	503	0	780	0	0	

Month	Regular Time Cost	Battery Cost	DVD Drive Cost	Material Cost	Penalty to BVE	Penalty to Pop	Revenue from BVE	Revenue from Pop	Laptop Holding Cost	Battery Holding Cost	DVD Drive Holding Cost	Overtime Cost	Hiring Cost	Layoff Cost	Shipping Cost
January	$52,500	$520,650	$420,070	$2,328,800	$0	$0	$816,000	$4,080,680	$37,100	$20,396	$12,360	$3,108	$35,000	$0	$17,002
February	$52,500	$280,070	$420,070	$2,423,600	$0	$0	$762,280	$3,500,000	$35,295	$8,262	$12,244	$6,426	$0	$0	$15,000
March	$52,500	$420,000	$420,070	$2,428,800	$0	$0	$720,120	$3,500,000	$35,360	$12,116	$12,102	$6,608	$0	$0	$15,000
April	$52,500	$420,000	$420,070	$2,420,400	$0	$0	$728,960	$3,500,000	$35,255	$12,014	$12,002	$6,314	$0	$0	$15,000
May	$57,000	$504,000	$504,070	$2,880,400	$0	$0	$760,920	$3,500,000	$40,325	$14,412	$14,402	$15,694	$3,000	$0	$15,000
June	$57,000	$504,000	$504,070	$2,880,400	$0	$0	$852,720	$4,080,680	$39,785	$14,410	$14,402	$15,694	$0	$0	$17,002
July	$52,500	$420,000	$420,070	$2,400,400	$0	$0	$769,760	$4,080,680	$34,465	$12,008	$12,002	$5,614	$0	$6,000	$17,002
August	$52,500	$420,000	$420,070	$2,400,400	$0	$0	$690,900	$3,500,000	$34,605	$12,006	$12,002	$5,614	$0	$0	$15,000
September	$52,500	$420,000	$420,070	$2,400,400	$0	$0	$693,700	$3,500,000	$34,705	$12,004	$12,002	$5,614	$0	$0	$15,000
October	$63,000	$539,500	$581,000	$3,320,000	$0	$0	$744,600	$3,500,000	$45,260	$16,604	$16,600	$22,120	$7,000	$0	$15,000
November	$63,000	$539,500	$581,000	$3,320,000	$0	$0	$900,320	$4,760,000	$45,020	$16,604	$16,600	$22,120	$0	$0	$19,000
December	$63,000	$520,065	$525,000	$3,000,000	$0	$0	$1,020,000	$4,760,000	$39,020	$17,008	$15,000	$10,920	$0	$0	$19,000
Revenue			$55,722,320												
Inventory			$5,197,795												
Penalty Received			$0												
Material and Component Cost			$43,347,015												
Labor Cost			$847,346												
Inventory Holding Cost			$785,757												
Shipping Cost			$194,006												
Penalty			$0												
Operating Expense			$3,600,000												
Income			$12,145,991												
Profit Margin			21.80%												

Cell	Formula	Note
M3	=IF(E3>B3*160,E3-B3*160)	Drag down to M14
N3	=IF(B3>B2,B3-B2,0)	Drag down to N14
O3	=IF(B3<B2,B2-B3,0)	Drag down to O14
B16	=B3*1500	Drag down to B27
C16	=IF(C3>=8001,C3*65,IF(C3>=4001,C3*70,C3*75))	Drag down to C27
D16	=IF(D3>=6001,D3*70,IF(D3>=3001,D3*72,D3*75))	Drag down to D27
E16	=400*E3	Drag down to E27
F16	=IF(F3>H3,(F3-H3)*50,0)	Drag down to F27
G16	=IF(G3>I3,(G3-I3)*100,0)	Drag down to G27
H16	=IF(H3>=2001,H3*675,IF(H3>=1001,H3*680,H3*700))	Drag down to H27
I16	=IF(I3>=6001,I3*680,IF(I3>=4001,I3*700,I3*750))	Drag down to I27
J16	=(J2-H3-I3+J3)/2*10	Drag down to J27
K16	=(C3+K3)/2*4	Drag down to K27
L16	=(D3+L3)/2*4	Drag down to L27
M16	=M3*14	Drag down to M27
N16	=N3*1000	Drag down to N27
O16	=O3*2000	Drag down to O27
P16	=5000+I3*2	Drag down to P27
D28	=SUM(H16:I27)	
D29	=J14*675+K14*65+L14*70	
D31	=SUM(C16:E27)	
D32	=SUM(B16:B27)+SUM(M16:O27)	
D33	=SUM(J16:L27)	
D34	=SUM(P16:P27)	
D35	=SUM(F16:G27)	
D36	=300000*12	
D37	=SUM(D28:D30)-SUM(D31:D36)	
D38	=D37/D28	

FIGURE 7.4

Profit calculation for Bell Inc.

We now show how to calculate profit for the retailers. We use BVE as an example. For Bell laptops, we use the actual order shown in Figure 7.2 and the actual receipt from Figure 7.3. For Ultra laptops, we assume that Ultra Corp. fully satisfies the actual order shown in Figure 7.2. The profit for BE is calculated as shown in Figure 7.5. The inventory holding cost is calculated by

	Actual Demand		Acutal Order		Actual Receipt		Actual Sales		Inventory	
Month	Bell Laptop	Ultra Laptop	Bell Laptop	Ultra Laptop	Bell Laptop	Ultra Laptop	Bell Laptop	Ultra Laptop	Bell Laptop	Ultra Laptop
									0	0
January	1120	2955	1200	4001	1300	4001	1120	2955	80	1046
February	1059	3611	1121	3000	1121	3000	1059	3611	142	435
March	1072	2278	1059	3000	1059	3000	1072	2278	129	1157
April	1119	2644	1072	3000	1072	3000	1119	2644	82	1513
May	1055	3347	1119	3000	1119	3000	1055	3347	146	1166
June	1132	2305	1254	3000	1254	3000	1132	2305	268	1861
July	1186	2575	1132	3000	1132	3000	1186	2575	214	2286
August	991	3179	987	3000	987	3000	991	3179	210	2107
September	1095	3550	991	3000	991	3000	1095	3550	106	1557
October	1025	3603	1095	3000	1095	3000	1025	3603	176	954
November	1539	3258	1324	3000	1324	3000	1500	3258	0	696
December	1348	2754	1500	3000	1500	3000	1348	2754	152	942

	Penalty Received		Product Cost		Revenue		Inventory Holding Cost		Shiping Cost
Month	Bell Laptop	Ultra Laptop	Bell Laptop	Ultra Laptop	Bell Laptop	Ultra Laptop	Bell Laptop	Ultra Laptop	
January	$0	$0	$816,000	$3,800,950	$1,120,000	$5,023,500	$7,680	$37,853	$15,402
February	$0	$0	$762,280	$3,000,000	$1,059,000	$6,138,700	$8,058	$33,608	$13,242
March	$0	$0	$720,120	$3,000,000	$1,072,000	$3,872,600	$7,980	$34,440	$13,118
April	$0	$0	$728,960	$3,000,000	$1,119,000	$4,494,800	$7,698	$42,525	$13,144
May	$0	$0	$760,920	$3,000,000	$1,055,000	$5,689,900	$8,082	$42,593	$13,258
June	$0	$0	$852,720	$3,000,000	$1,132,000	$3,918,500	$10,008	$45,203	$13,508
July	$0	$0	$769,760	$3,000,000	$1,186,000	$4,377,500	$9,684	$53,603	$13,264
August	$0	$0	$890,900	$3,000,000	$991,000	$5,404,300	$8,456	$55,448	$12,974
September	$0	$0	$693,700	$3,000,000	$1,095,000	$6,035,000	$7,842	$49,980	$12,982
October	$0	$0	$744,600	$3,000,000	$1,025,000	$6,125,100	$8,262	$41,333	$13,190
November	$0	$0	$900,320	$3,000,000	$1,500,000	$5,538,600	$9,000	$34,875	$13,648
December	$0	$0	$1,020,000	$3,000,000	$1,348,000	$4,681,800	$9,912	$34,785	$14,000

Revenue	$75,002,300
Inventory	$1,045,360
Penalty Received	$0
Product Cost	$46,261,230
Inventory Holding Cost	$608,915
Shiping Cost	$161,710
Operating Expense	$18,000,000
Income	$11,015,806
Profit Margin	14.69%

Cell	Formula	Note
H4	=IF(J3+F4>B4,B4,J3+F4)	Drag down to H15
I4	=IF(K3+G4>C4,C4,K3+G4)	Drag down to I15
J4	=J3+F4-H4	Drag down to J15
K4	=K3+G4-I4	Drag down to K15
D18	=IF(D4>F4,(D4-F4)*50,0)	Drag down to D29
E18	=IF(E4>G4,(E4-D4)*100,0)	Drag down to E29
F18	=IF(D4>=2001,F4*675,IF(D4>=1001,F4*680,F4*700))	Drag down to F29
G18	=IF(E4>=4001,G4*950,IF(E4>=2001,G4*1000,G4*1100))	Drag down to G29
H18	=H4*1000	Drag down to H29
I18	=I4*1700	Drag down to I29
J18	=(F4+J3+J4)/2*12	Drag down to J29
K18	=(K3+G4+K4)/2*15	Drag down to K29
L18	=5000+F4*2+G4*2	Drag down to L29
D30	=SUM(H18:I29)	
D31	=J15*680+K15*1000	
D32	=SUM(D18:E29)	
D33	=SUM(F18:G29)	
D34	=SUM(J18:K29)	
D35	=SUM(L18:L29)	
D36	=1500000*12	
D37	=SUM(D30:D32)-SUM(D33:D36)	
D38	=D37/D30	

FIGURE 7.5
Profit calculation for BVE Store.

averaging the beginning inventory and the ending inventory in each month. The beginning inventory for laptops includes those purchased from the manufacturers. The value of the remaining inventory at the end of December (Cell D31) is calculated using the most recent purchase price.

After the simulation, students can study the operation of their companies and the supply chain to see how their decisions impact profitability. They are encouraged to discuss strategies to improve company/supply chain and the pros and cons of these strategies. For example, BVE was able to fully satisfy the demand of Ultra laptops but not that of Bell laptops. Should BVE increase the safety inventory level for Bell laptops? A laptop manufacturer may find that it cannot meet its production goal due to the shortage of components. What can it do to solve this problem? Should it hold a higher level of component safety inventory? Or should it restructure the sales contract to impose a steeper penalty on the component supplier? This simulation game and the subsequent analysis can help students synthesize the knowledge acquired and improve their supply chain operation skills.

7.4 Notes for Using the Simulation Game

It is recommended that the instructor uses this simulation game as a class project. The simulation game requires the participation of eight companies with different levels of workload. Both BVE and Pop Electronics need to negotiate two sales contracts and determine their ordering strategies. Both Bell and Ultra need to negotiate at least four and as much as six sales contracts, determine their ordering strategies, and develop their production plans. Longey, Everlast, Alpha, and Beta need to negotiate one or two sales contracts, develop their production plan, and determine their safety inventory strategies. Therefore, BVE and Pop Electronics can each be adequately managed by a single student; Bell and Ultra should be managed by two to three students; whereas Longey, Everlast, Alpha, and Beta can be managed by one or two students. For a large class, multiple supply chains can be formed, each involving the same eight companies and using the same data.

This class project requires considerable amount of time to complete due to the need for contract negotiation. It is recommended that the project to be assigned 6–8 weeks prior to its due date, preferably after the subjects of forecasting and aggregate planning are taught. Students should have sufficient knowledge to complete the project after inventory management is also taught. Although the students will work on the project outside the classroom, a certain amount of class time should be reserved for face-to-face contract negotiation. It is recommended that two such class sessions are reserved. The first one should be held 1 or 2 weeks after the project is assigned, where the students become familiar with all the players involved and their potential suppliers/customers. The second one should be held 2 to 3 weeks before the project is due, where the students aim to reach a consensus with their suppliers/customers. They can then finalize the sales contracts outside the classroom.

After a company finalizes all the sales contracts and makes all relevant decisions, it is ready to calculate its profit. Note that the calculation can only be done based on the sales contracts negotiated by the company and its production plan (in the case of laptop manufacturers and component suppliers) and assuming that the demand forecast is accurate (in the case of retailers). The result will be different from that based on the "actual" demand (given by the instructor), not only because of the difference in forecasted and "actual" demand but also because of the interactions among the companies (e.g., order may not be fully satisfied). An Excel spreadsheet tool is provided to simulate the actual operation of the supply chain. Once all the sales contracts and production plans are entered into the spreadsheet and the "actual" demand is created, the tool will perform all the calculations automatically. It is recommended that the instructor require the students to turn in their sales contracts, production plans, and ordering/inventory strategies, along with their estimated profits. The instructor then enters the required information in the Excel spreadsheet tool and provides the results to the students. Students are then required to review the "actual" results and provide reports summarizing what they have learned. A debriefing session can then be held by the instructor to discuss project outcomes and lessons learned.

Index